MATH HOUSE Online

Learn By Doing: Algebra Essentials

Repairing Your Algebra Foundation

Ryan Hobbs

Delaware
Spring Press

ISBN: 978-1-7332514-7-1

The traditional lecture videos, linked by QR code at the end of each lesson, use material from Marecek, Lynn, et al. *Elementary Algebra 2e*. OpenStax, 2020, which is available for free at https://openstax.org/details/books/elementary-algebra-2e. The traditional lecture videos are available for free at https://www.youtube.com/@rphobbs2002.

This book is built from excerpts out Learn by Doing: Algebra I.

For additional help, see our complete line of tutoring resources:
Learn By Doing: Algebra I
Learn By Doing: Algebra II
Learn By Doing: Trigonometry
Learn By Doing: PreCalculus

The Reason for this Workbook

I have been a college math lecturer for over ten years and it didn't take me long to develop my foremost principle of education—if students aren't awake then they aren't learning. In my opinion, for most math students, the traditional lecture approach has limited value. And so, I began a journey to develop a more interactive approach to the mathematics classroom. This workbook is the result.

However, there is much more here than just a series of worksheets, as I've also strived to incorporate the other principles of math education which I've discovered over the years.

Math is a language. Good teaching must translate this language of numbers into English. Language learning also requires "comprehensible input." It is good to be challenged and stretched, but when teaching is beyond a student's level of understanding they become overwhelmed and shut down. As math progresses, it is valuable to show students how more advance concepts relate to simpler concepts which they already know and understand. And, when possible, it helps to make math visual.

Yet perhaps most importantly, good teaching comes beside students when they need help. This is why every activity has a QR code which links to a walk-through video of the active learning lesson.

In other words, good teaching is very much like good tutoring, and that is the fundamental idea which has guided the development of this course.

How to Use this Book

Each activity in this workbook has a QR code in the top righthand corner. This QR code will take you to a YouTube video where I work and explain every problem and concept in that lesson. It is my attempt to use technology to come along side of a student just as I would do in a real classroom.

So, as you work the activity, start and stop the video, as needed, for help. Or, skip through any part of the video which you don't need. The end of each lesson also contains a solution guide for checking your answers.

If after finishing an activity, you feel as if you could use some additional help, a QR code in the bottom right links to a traditional lecture for that section.

Table of Contents

Introduction

For over ten years, I've been trying to help college students who struggle with Algebra. And piece by piece, the concepts in this workbook have been developed. There are many reasons why students find themselves struggling with Algebra, but there are two which run through nearly everyone I encounter. First, Algebra doesn't make sense. Second, they were pushed on too quickly without understanding the essentials.

You can't build on a shaky foundation. And you can't learn higher level math when the key concepts of Algebra aren't strong. I designed this workbook to be foundation repair for students who are having difficulty with a College Algebra course. However, the ideas here will help anyone who is struggling with Algebra regardless of level.

As teachers, I've come to realize that we don't spend enough time answering the question, "What is Algebra?" I think we fail to do this because it isn't all that easy to answer. Ten years into college lecturing and I believe I've finally found a simple way to explain it. To answer it, I'll begin by answering the logical question which comes before it, "What is Mathematics?"

Math is a language, just like English or Spanish or French. But unlike those other languages which exist to allow people to communicate together, math exists to communicate numbers. The real world is full of numbers. And while we can explain numbers in English (or Spanish or French), math is a language that communicates numbers concisely. Although it isn't as common as it once was, an appropriate comparison to mathematics is shorthand. Shorthand is a system of abbreviations and symbols used to record language more quickly. Whole sentences can be summed by a few strange squiggles and curves. Similarly, a few abbreviations and symbols in math can record numbers in a way that would require sentences to explain in English.

So then, what is Algebra? Well, Algebra is a smaller part of the language of mathematics where we are representing a number which we don't know with a letter, called a variable. In other words, the dreaded x. And, I've come to realize that much, if not all, of Algebra is based on three ideas: solving equations, factoring, and relations. These three concepts, along with some foundational skills, will provide the structure for the sections to this workbook.

Each of the four sections of the workbook begins with an explanation of that main idea's importance to Algebra. Then, a series of interactive lessons will walk you through the concepts using the methods which I've found to be the most effective. If you need help, each lesson includes a QR code which links to a help video where I walk through the entire lesson. Finally, additional practice problems, and their solutions, have been provided.

So, that's how it works. Now, let's get started repairing your foundation and learning the essentials of Algebra.

The Foundation

This section is filled with the most essential skills to succeed in Algebra. Over the past decade, when I see a student trying to climb the math ladder without having mastered the ideas you find here, I know it is going to be a struggle. When I talk about repairing your math foundation this is absolutely where to begin. Students have seen these ideas before and often have a difficult time admitting to me that they don't know them. However, I find that pride is one of the greatest barriers to math success for struggling learners. Even if these activities seem easy to you, take your time and be certain that you have them. And, as you move forward in the workbook, should you need to, return and review these ideas again.

Lesson 1: Prime and Composite Numbers; Factor Trees

I believe that understanding factoring is one of the keys to algebra. And so, we want to begin by examining how numbers can be "composed" of other numbers.

First, numbers can be classified into two categories: prime numbers and composite numbers. Composite numbers are made up of two or more other numbers which are being multiplied together.

$$6 = 2 \cdot 3$$

$$15 = 3 \cdot 5$$

$$30 = 2 \cdot 3 \cdot 5$$

Prime numbers are those which cannot. They can only be written as a product by multiplying themselves by 1.

$$5 = 5 \cdot 1$$

$$7 = 7 \cdot 1$$

$$59 = 59 \cdot 1$$

a) Look at this list of numbers. Circle any primes.

8   13   21   35   17   61   100

Finding the factors which make up a number is a very important algebra skill. In order to find factors, we need to know what numbers can divide into our number. To make that easier, there are divisibility rules. If the number we are dividing fits certain rules, we can discover factors more quickly.

Here is a list of the most common divisibility rules:

- Divisible by 2: If the number is an even number. (The last digit is a 0, 2, 4, 6, or 8.)
- Divisible by 3: If the sum of the digits is divisible by 3.
- Divisible by 5: If the last digit ends in 5 or 0.
- Divisible by 6: If a number is divisible by 2 and 3.
- Divisible by 10: If a number ends in 0.

b) Circle any numbers divisible by 2.

15   16   102   2434   4   16   81   156

c) Circle any number divisible by 3.

For example: 132. Add the digits. $1 + 3 + 2 = 6$. Six is divisible by 3, so 132 is divisible by 3.

231   6033   122   42   63   99   10547

d) Circle any number divisible by 5.

15   103   7855   239670   65   8000   98

e) Circle any numbers divisible by 6.

For example: 462. It is divisible by 2 because it ends in an even number. It is divisible by 3 because $4 + 6 + 2 = 12$ is divisible by 3.

81   240   550   1212   142   6954

f) Circle any numbers divisible by 10.

1000   5020   5025   100365   84750   90   21054780

In algebra, it will often be useful to find the prime factorization of a number. If a number is composite, it is the product of other numbers. For instance, here is the prime factorization of the number 240.

$$240 = 2 \cdot 2 \cdot 2 \cdot 2 \cdot 3 \cdot 5$$

It is valuable to understand that the number 240 is the same as $2 \cdot 2 \cdot 2 \cdot 2 \cdot 3 \cdot 5$. Finding a prime factorization requires dividing down 240. I use a method called a factor tree.

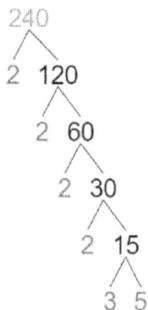

240
2 120
    2 60
        2 30
            2 15
                3 5

Keep dividing the number. When a number is prime—stop.

The prime factorization is the list of all the prime number multiplied together.

$$240 = 2 \cdot 2 \cdot 2 \cdot 2 \cdot 3 \cdot 5$$

Using the division rules is helpful, but I usually just try to divide by 2. If that doesn't work, I try 3. Then, I try 5. Of course, there are more possibilities, but most numbers will depend on those three.

Here is the factor tree for 165:

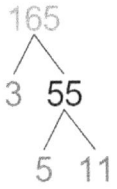

```
    165
    /\
   3  55
      /\
     5  11
```

g) Write the prime factorization.

165:

Here is the factor tree for 2358:

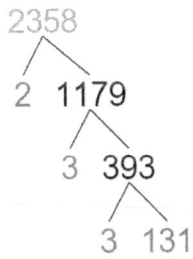

```
    2358
    /\
   2  1179
       /\
      3  393
          /\
         3  131
```

h) Write the prime factorization.

2358:

i) Make the factor tree and give the prime factorization for 81.

j) Make the factor tree and give the prime factorization for 64.

k) Make the factor tree and give the prime factorization for 125.

One of the applications of prime factorizations is to find something called the least common multiple (LCM) for a pair of numbers. First, let's see the most basic way to find the LCM. Suppose we wanted to find the LCM for the numbers 8 and 10. We could, literally, write out all the multiples for each number until we found the lowest number that they both share.

$$8: 8, 16, 24, 32, 40, 48, 56, 64, 72, 80 \ldots$$

$$10: 10, 20, 30, 40, 50, 60, 70, 80 \ldots$$

If we compare the lists, we find that the lowest match if 40. Therefore, the LCM is 40. Notice that there are other multiples. 80 is also a match. However, it is not the lowest.

Here is a list of the multiples for the numbers 12 and 15.

$$12: 12, 24, 36, 48, 60, 72, 84, 96 \ldots$$

$$15: 15, 30, 45, 60, 75, 90 \ldots$$

l) What is the least common multiple?

m) Try this list approach to find the LCM for the numbers 15 and 21.

15:

21:

The list method works, but it can be time consuming, especially with larger numbers. Prime factorizations can make the work much easier.

8

Here are the prime factorizations for 8 and 10:

$8 = 2 \cdot 2 \cdot 2$

$10 = 2 \cdot 5$

I've circled factors which they share.

$$8 = \boxed{2}\ 2 \cdot 2$$
$$10 = 2\ 5$$

Now, create a factorization with every factor from both numbers. However, count the shared factors only once.

$2 \cdot 2 \cdot 2 \cdot 5$

The least common multiple is: $2 \cdot 2 \cdot 2 \cdot 5 = 40$

Let's walk through the idea again. This time with 12 and 30.

Here are the prime factorizations.

$12 = 2 \cdot 2 \cdot 3$

$30 = 2 \cdot 5 \cdot 3$

Now, we circle matching factors.

$$12 = 2\ 2\ 3$$
$$30 = 2\ 5\ 3$$

Counting matches only once, we list out the factors to find the LCM.

$$2 \cdot 2 \cdot 5 \cdot 3 = 60$$

Use the prime factorization method to find the LCM for the following:

n)   Find the LCM of 12 and 15.

o) Find the LCM for 15 and 21.

p) Find the LCM for 192 and 168.

a) Look at this list of numbers. Circle any primes:

21 8 19 53 26 65 61 57

b) Circle any numbers divisible by 2:

3 14 21 106 81 90 63 1256

c) Circle any numbers divisible by 3:

11 33 96 14 41 54 117 2169

d) Circle any numbers divisible by 5:

10 41 45 506 704 8005 125680

e) Circle any numbers divisible by 6:

21 44 30 108 174 56 165 150

f) Make the factor tree and give the prime factorization for the number 105.

g) Make the factor tree and give the prime factorization for the number 162.

h) Make the factor tree and give the prime factorization for the number 250.

i) Use the prime factorization approach to find the LCM of the following numbers: 60 and 42.

j) Use the prime factorization approach to find the LCM of the following numbers: 297 and 330.

Lesson One: Answers

a) 13, 17, 61

b) 16, 102, 2434, 4, 16, 156

c) 231, 6033, 42, 63, 99

d) 15, 7855, 239670, 65, 8000

e) 240, 1212, 6954

f) 1000, 5020, 84750, 90, 21054780

g) 165: 3 • 5 • 11

h) 2358: 2 • 3 • 3 • 131

i) 81: 3 • 3 • 3 • 3

j) 64: 2 • 2 • 2 • 2 • 2 • 2

k) 125: 5 • 5 • 5

l) 60

m) 105

n) 3 • 2 • 2 • 5 = 60

o) 3 • 5 • 7 = 105

p) 2 • 2 • 2 • 2 • 2 • 2 • 3 • 7 = 1344

Lesson One: Additional Exercises Answers

a) 19, 53, 61

b) 14, 106, 90, 256

c) 33, 96, 54, 117, 2169

d) 10, 45, 8005, 125680

e) 30, 108, 174, 150

f) 105 = 3 • 5 • 7

g) 162 = 2 • 3 • 3 • 3 • 3

h) $250 = 2 \cdot 5 \cdot 5 \cdot 5$

i) $2 \cdot 2 \cdot 3 \cdot 5 \cdot 7 = 420$

j) $2 \cdot 3 \cdot 3 \cdot 3 \cdot 5 \cdot 11 = 2970$

Lesson 2: Adding and Subtracting Positive and Negative Numbers

Adding and subtracting negative numbers can give people a lot of difficulty. There are various ways to try and teach the idea. I suggest thinking of it like money. A negative sign means debt.

$$-10 + 8$$

You owe 10 dollars and then you get 8. Your debt is now -2.

$$-10 + 8 = -2$$

Try these:

a) $-7 + 5 =$

b) $-21 + 7 =$

In these next problems, you will get out of debt:

c) $-5 + 8 =$

d) $-21 + 30 =$

e) $-1 + 15 =$

Look at this problem:

$$-10 - 8$$

You owe 10 dollars and then you owe 8 more. Your debt is now -18.

$$-10 - 8 = -18$$

Work these:

f) $-11 - 9 =$

g) $-8 - 3 =$

h) $-5 - 7 =$

Next, I've mixed them up. Work these. Remember, think of it as debt.

i) $-19 - 2 =$

j) $-1 + 18 =$

k) $14 - 7 =$

l) $-16 + 8 =$

m) $-5 - 6 =$

Secondly, we never want there to be more than one addition or subtraction sign in a row. So, when that occurs, we'll begin with a fix. The key is to think of negative signs as opposites. Addition signs don't change anything. Therefore:

$$- + \text{ :The opposite of a positive is: } -$$

$$+ - \text{ :The positive doesn't change the negative so: } -$$

$$- - \text{ :The opposite of an opposite: } +$$

$$+ + \text{ :Two positives don't change anything: } +$$

That's the logic, but here's the summary:

$$- + \text{ will be } -$$

$$+ - \text{ will be } -$$

$$- - \text{ will be } +$$

$$+ + \text{ will be } +$$

Simply the following. Turn the two signs into one.

n) $-+5 =$

o) $--14 =$

p) $+-2 =$

q) $-+9 =$

r) $++6 =$

In these next problems, the parentheses don't change anything. So, ignore the parentheses and simplify the two signs in a row.

s) $-(+8) =$

t) $+(+10) =$

u) $-(-12) =$

v) $+(-1) =$

Now, simplify the following by removing the double signs. For instance:

$$8 + (-6) = 8 - 6 = 2$$

w) $17 - (-2) =$

x) $10 - (+7) =$

y) $6 + (+5) =$

z) $-6 - (-7) =$

aa) $-41 + (-3) =$

bb) $-20 - (+9) =$

cc) $17 + (-11) =$

Finally, follow the order of operations to simplify these. Whenever you get two signs in a row, use the ideas that we've examined above.

$$9 + (-5 - 1) = 9 + (-6)$$
$$9 - 6 = 3$$

Work these:

dd) $22 + (-11 - 3) =$

ee) $-5 + (-19 - 2) =$

ff) $17 - (3 + 13) =$

gg) $-2 - (-12 + 6) =$

hh) $-40 - (-25 - 5) =$

ii) $20 - (-12 + 8) - 3 =$

Lesson Two: Additional Exercises

Simplify the following:

a) $-4 + 16$

b) $8 - 20$

c) $-1 - 13$

d) $-1 + 13$

e) $- - 7$

f) $- + 34$

g) $+(-19)$

h) $-(-4)$

i) $-32 - (+14)$

j) $17 - (-17)$

k) $-25 + (-5)$

l) $11 + (-9 - 3)$

m) $-5 - (16 - 20)$

Lesson Two: Answers

a) -2

b) -14

c) 3

d) 9

e) 14

f) -20

g) -11

h) -12

i) -21

j) 17

k) 7

l) -8

m) -11

n) -5

o) 14

p) -2

q) -9

r) 6

s) -8

t) 10

u) 12

v) -1

w) 19

x) 3

y) 11

z) 1

aa) -44

bb) -29

cc) 6

dd) 8

ee) -26

ff) 1

gg) 4

hh) -10

ii) 21

Lesson Two: Additional Exercises Answers

a) 12

b) -12

c) -14

d) 12

e) 7

f) -34

g) -19

h) 4

i) -46

j) 34

k) -30

l) -1

m) -1

Foundations

Lesson 3: Multiplying Positive and Negative Numbers

Multiply the following:

a) $2 \cdot 5 =$

b) $5 \cdot 2 =$

c) $1 \cdot 7 =$

d) $7 \cdot 1 =$

e) $3 \cdot 2 \cdot 1 =$

f) $1 \cdot 3 \cdot 2 =$

g) $2 \cdot 3 \cdot 1 =$

h) Does the order in which we multiply matter?

We looked at adding and subtracting positive and negative numbers. Now we want to look at multiplying and dividing them. There is a pattern that can be memorized, but let's examine where it comes from.

Let's multiply:

$$-5 \cdot -3$$

The signs can be thought of like numbers, and since the order we multiply numbers doesn't matter, we can move the negatives to the front.

$$--5 \cdot 3 =$$

If we multiply the numbers then we get 15. And we learned in the last lesson that two negatives in a row is a positive. So:

$$--5 \bullet 3 = +15$$

For these problems, show how to move the negatives to the front. You won't normally have to, but I want you to get comfortable with where the final sign comes from.

i) $\quad -3 \bullet -8 =$

j) $\quad -9 \bullet -7 =$

k) $\quad -6 \bullet -11 =$

What will the final sign always be if we multiply two negative numbers together?

Next, let's multiply positive and negative numbers.

$$-6 \bullet 3$$

Move the signs to the front.

$$- +6 \bullet 3$$

The numbers make 18. We learned previously that a negative multiplied by a positive gives a negative.

$$- +6 \bullet 3 = -18$$

Try some. Again, show the signs moving to the front.

l) $\quad -11 \bullet 5 =$

m) $\quad 6 \bullet -7 =$

n) $\quad -8 \bullet 3 =$

o) $\quad 2 \bullet -4 =$

p) What will the final sign always be if we multiply a positive number by a negative number?

The same ideas carry over to division. Move the signs to the front.

$$\frac{-15}{5}$$

$$-+\frac{15}{5}$$

15 divided by 5 is 3. And, a negative times a positive makes a negative. So:

$$-+\frac{15}{5} = -3$$

Divide the following:

q) $\dfrac{-27}{3} =$

r) $\dfrac{45}{-9} =$

s) $\dfrac{-42}{-7} =$

t) $-55 \div 11 =$

u) $-81 \div -9 =$

Next, let's look at exponents.

$$(-2)^4$$

This is really the same as:

$$-2 \cdot -2 \cdot -2 \cdot -2$$

Let's move the negatives out front.

$$----2 \cdot 2 \cdot 2 \cdot 2$$

Two negatives in a row make a positive.

$$\overline{(--}\ \overline{)(-}\ \overline{)}2 \cdot 2 \cdot 2 \cdot 2$$

So:

$$++2 \cdot 2 \cdot 2 \cdot 2 = 16$$

v) Try one. Move the negatives out front and then simplify.

$$(-2)^6$$

w) What will the final sign be any time you have an even exponent?

Let's work an odd exponent.

$$(-2)^3$$

$$-2 \cdot -2 \cdot -2 =$$

I've moved the negatives out front.

$$\overline{(--}\ \overline{)}-2 \cdot 2 \cdot 2 =$$

The first two make a positive.

$$\overline{(+-}\ \overline{)}2 \cdot 2 \cdot 2 =$$

A positive and a negative make a negative.

$$-2 \cdot 2 \cdot 2 = -8$$

Try one. Move the exponents out front.

x) $(-2)^5 =$

24

y)   What will the final sign be any time you have an odd exponent?

Work these problems:

z) $(-3)^3 =$

aa) $(-5)^2 =$

bb) $(-4)^4 =$

cc) $(-6)^3 =$

Next, we will simplify expressions involving negative numbers. Remember to follow the order of operations.

dd) $-2(5-8)+(-3)^2$

ee) $20 \div 2 - 3(10-7) + 4(2-6) + (-2)^3$

Finally, we will evaluate expressions. Evaluating means we will replace the variable with a value that is given to us. When we do this, we put the value in parenthesis to make sure we don't miss out on any potential multiplication.

Evaluate $x + 4$ when $x = -7$.

$(-7) + 4 =$

There is no multiplication here, so we can simply drop the parenthesis.

ff) $-7 + 4 =$

In each of these problems, after replacing the variable, follow the order of operations.

gg)   Evaluate $-(v - 8)$ when $v = -3$.

hh)   Evaluate $x^2 - 3x + 15$ when $x = -2$.

ii)   Evaluate $-2x^2 + 6x - 10$ when $x = -1$.

These last two problems have two variables.

jj)   Evaluate $(x - y)^2$ when $x = 3$ and $y = -2$.

kk)   Evaluate $(x + y)^3$ when $x = -2$ and $y = -3$.

Simplify the following:

a) $-3 \cdot 11$

b) $-8 \cdot -4$

c) $15 \cdot -2$

d) $-6 \cdot -6$

e) $\dfrac{-56}{8}$

f) $\dfrac{-55}{-11}$

g) $-63 \div 7$

h) $42 \div -6$

i) $(-3)^4$

j) $(-2)^7$

k) $-3(4-8)+(-4)^2$

l) $5 \cdot 6 - (-5)^3 + (125 \div 25) + (-2)^4$

m) Evaluate $-(x+15)$ when $x = -8$.

n) Evaluate $2y^2 - 5y + 16$ when $y = -3$.

o) Evaluate $x^3 - y$ when $x = -2$ and $y = -8$.

p) Evaluate $-5(x - y)$ when $x = -4$ and $y = 2$.

Lesson Three: Answers

a) 10

b) 10

c) 7

d) 7

e) 6

f) 6

g) 6

h) No

i) 24

j) 63

k) 66

l) positive; -55

m) -42

n) -24

o) -8

p) negative

q) -9

r) -5

s) 6

t) -5

u) 9

v) 64

w) positive

x) -32

y) negative

z) -27

aa) 25

bb) 256

cc) -216

dd) 15

ee) -23

ff) −3

gg) 11

hh) 25

ii) -18

jj) 25

kk) -125

Lesson Three: Additional Exercises Answers

a) -33

b) 32

c) -30

d) 36

e) -7

f) 5

g) -9

h) -7

i) 81

j) -128

k) 28

l) 176

m) -7

n) 49

o) 0

p) 30

Foundations

Lesson 4: Multiplying and Dividing Fractions

In this activity, we are going to refresh your skills with fractions. It seems as if no one enjoys working with fractions, but the concepts here are very important as we move into algebra. All of the ideas will reappear frequently, and the better that you understand them here, the easier the algebraic version will become.

Equivalent Fractions

Fractions are like slices of a pie. Here is the fraction $\frac{1}{3}$.

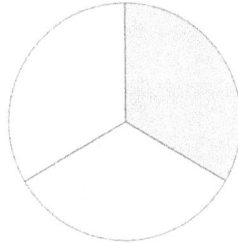

The bottom of the fraction is called the denominator, and it tells us how we have sliced the pie. With $\frac{1}{3}$, we have sliced the pie into three pieces. The top of the fraction is called the numerator, and it tells us how many slices we have. With $\frac{1}{3}$, we have one of the three slices.

However, we could slice the same pie differently. This is $\frac{2}{6}$.

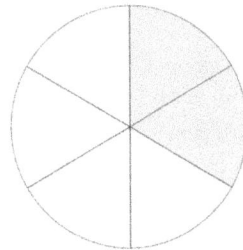

Although there are now six slices of pie, notice that our portion is still the same.

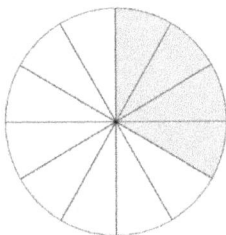

a) I've cut the pie into another equivalent fraction. What is the fraction?

31

In each case, the fractions are the same.

$$\frac{1}{3} = \frac{2}{6}$$

But, how do we find equivalent fractions. These images tell us.

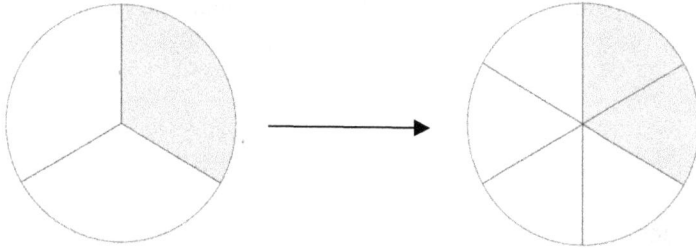

We doubled the number of slices. And that double the amount of slice which are shaded.

$$\frac{1 \cdot 2}{3 \cdot 2} = \frac{2}{6}$$

It is like an equation. If we double the bottom, we must double the top. There are any number of possible equivalent fractions. For instance, we could have tripled the slices. Finish the equivalent

b) fraction:

$$\frac{1 \cdot 3}{3 \cdot 3} =$$

c) If we start with $\frac{2}{5}$, which of the following would be an equivalent fraction?

$$\frac{8}{15}$$

$$\frac{6}{15}$$

$$\frac{6}{20}$$

$$\frac{12}{20}$$

d) Give three equivalent fractions for $\frac{2}{3}$?

32

We can, of course, go the other direction.

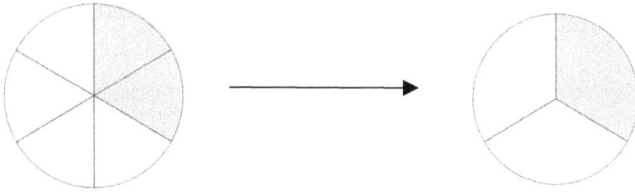

Here, we have cut half as many slices, so we have half the number of shaded pieces. But to do this without a picture requires factoring. Turn the numerator and denominator into their prime factorization.

$$\frac{2}{6} = \frac{2}{2 \cdot 3}$$

If you have any matching factors in the numerator and denominator, cancel them.

$$\frac{2}{6} = \frac{\cancel{2}}{\cancel{2} \cdot 3} = \frac{1}{3}$$

Notice that we don't cancel the numerator down to nothing. Reducing leaves a 1 behind.

e) Select the correct reduced fraction. $\frac{15}{30} = \frac{3 \cdot 5}{2 \cdot 3 \cdot 5}$

$\frac{3}{10}$

$\frac{1}{15}$

$\frac{1}{3}$

$\frac{1}{2}$

A prime factorization tree is a helpful tool for fractions. Slice a number down to its prime numbers.

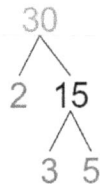

For even numbers, I keep dividing by 2. If two doesn't work, I try 3. Or if it ends in zero or five, I divide by 5. There are, of course, other possibilities, but these will typically help you get started. Notice on the right that the 2 was prime but 15 wasn't. So, I divided the 15 again.

f) Reduce the following fractions. I've given you the factor tree for the first two problems.

$$\frac{21}{49}$$

g) After you reduce, you'll have two factors left in the denominator. Simply multiply those factors back together.

$$\frac{18}{24}$$

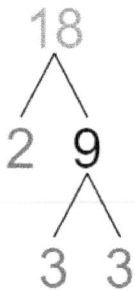 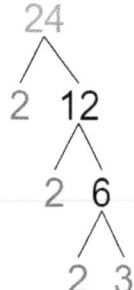

h)

$$\frac{40}{48}$$

i) This next problem has a negative out front. Just work the fraction like normal and return the negative to the final fraction.

$$-\frac{45}{54}$$

j) Try one more. This time the problem involves large numbers. However, the process remains the same. Use your tree to create the prime factorization for the numerator and denominator and cancel any matches.

$$-\frac{96}{144}$$

We can also extend the idea to algebra.

$$\frac{7x}{7y}$$

The x and the y are just factors.

$$\frac{7 \cdot x}{7 \cdot y}$$

We cancel any matching factors and get:

$$\frac{7 \cdot x}{7 \cdot y} = \frac{x}{y}$$

Work these.

k) $\dfrac{2a}{2b} =$

l) $\dfrac{3x}{6y} =$

Multiplying Fractions

Multiply the following:

m) $8 \cdot \dfrac{1}{2} =$

n) $20 \cdot \dfrac{1}{2} =$

Multiplying by half means that we end with half of what we started with. Suppose we started with a fraction like $\frac{1}{4}$.

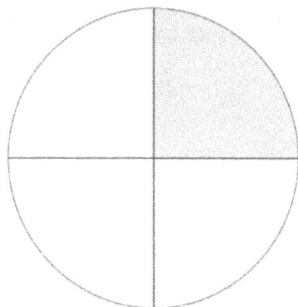

Multiplying by $\frac{1}{2}$ should give us half as much as when we started.

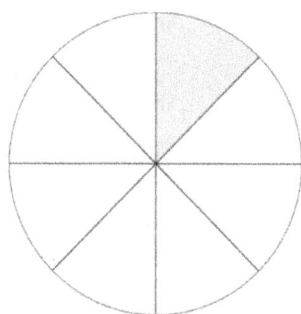

So, how does the math work.

$$\frac{1}{4} \cdot \frac{1}{2} =$$

We simply multiply across the top, and we multiply across the bottom.

$$\frac{1}{4} \cdot \frac{1}{2} = \frac{1}{8}$$

(Notice that this matches our expected slice of pie.)

Multiply the following fractions.

o) $\frac{1}{3} \cdot \frac{1}{2} =$

We can multiply by any fraction and it follows the same concept.

p) $\frac{1}{4} \cdot \frac{1}{5} =$

If you multiply across and your answer can be reduced, follow the procedure for simplifying fractions which we saw earlier.

q) $\dfrac{2}{3} \cdot \dfrac{1}{6} =$

r) $\dfrac{3}{10} \cdot \dfrac{4}{5} =$

There is a negative in this next problem. Just bring the negative out front.

s) $-\dfrac{6}{7} \cdot \dfrac{3}{8} =$

This problem has two negatives. Remember, if you multiply two negative numbers you get a positive.

t) $-\dfrac{5}{6} \cdot -\dfrac{1}{5} =$

Again, this can be extended to algebra. Just multiply across the tops and across the bottoms.

$$\dfrac{2}{3} \cdot \dfrac{x}{5} = \dfrac{2x}{15}$$

Try a couple:

u) $\dfrac{1}{7} \cdot \dfrac{y}{6} =$

This next problem can reduce.

v) $\dfrac{x}{10} \cdot \dfrac{2}{5} =$

Finally, let's look at division of fractions. Suppose I want to work the following problem:

$$\frac{1}{4} \div 2$$

This would mean that I have one-fourth of a slice and I want to share it among two people.

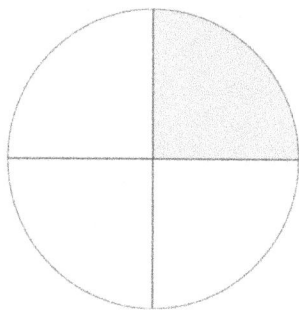

$$\div 2 =$$

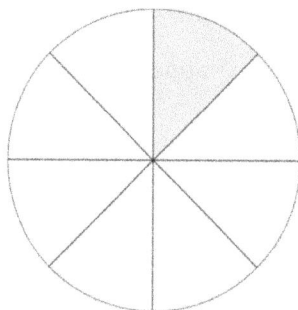

$$\frac{1}{4} \div 2 = \frac{1}{8}$$

Notice that this is the same answer we got when we multiplied by $\frac{1}{2}$.

$$\frac{1}{4} \div 2 = \frac{1}{4} \cdot \frac{1}{2} = \frac{1}{8}$$

Dividing by fractions is hard to wrap our mind around. However, what we need to see is that we get the same answer if we multiply by the number turned upside down (the reciprocal). This always works and creates a strategy for dividing fractions: Keep, Change, Flip.

Keep the first number.

Change the sign to multiplication.

Flip the last number upside down.

Let me work one.

$$\frac{1}{4} \div \frac{2}{3} =$$

Keep the $\frac{1}{4}$.

Change to multiplication.

Flip the $\frac{2}{3}$ upside down and make it $\frac{3}{2}$.

$$\frac{1}{4} \cdot \frac{3}{2} = \frac{3}{8}$$

Try some.

w) $\frac{1}{3} \div \frac{2}{5} =$

On your next problem, the final answer will reduce.

x) $\frac{3}{4} \div \frac{3}{5} =$

This problem involves a negative number. Just bring the negative over to the final answer.

y) $-\frac{4}{7} \div \frac{2}{5} =$

And this problem has two negatives. Remember, two negatives multiply to get a positive.

z) $-\frac{7}{8} \div -\frac{7}{4} =$

Work one with more difficult fractions. After multiplying across the top and bottom, reduce down.

aa) $\frac{27}{40} \div \frac{72}{55} =$

And, finally, the ideas can again be extended to algebra.

$$\frac{x}{4} \div \frac{1}{3} = \frac{x}{4} \cdot \frac{3}{1} = \frac{3x}{4}$$

Try these:

bb) $\quad \frac{1}{5} \div \frac{3}{7x} =$

cc) $\quad \frac{5}{4} \div \frac{2x}{3} =$

dd) $\quad \frac{y}{8} \div \frac{3}{12} =$

Lesson Four: Additional Exercises

a) Give three equivalent fractions for $\frac{3}{7}$.

Reduce the following fractions:

b) $\dfrac{8}{28}$

c) $\dfrac{18}{27}$

d) $\dfrac{35}{45}$

e) $-\dfrac{24}{60}$

f) $\dfrac{150}{225}$

g) $\dfrac{21x}{3y}$

Multiply the following fractions:

h) $\dfrac{1}{3} \cdot \dfrac{1}{5} =$

i) $\dfrac{2}{7} \cdot \dfrac{3}{7} =$

j) $\dfrac{4}{5} \cdot \dfrac{3}{6} =$

k) $-\dfrac{4}{9} \cdot \dfrac{3}{5} =$

l) $\dfrac{x}{3} \cdot \dfrac{4}{5} =$

m) $\dfrac{3}{10} \cdot \dfrac{2y}{9} =$

Divide the following fractions.

n) $\dfrac{1}{4} \div \dfrac{2}{3} =$

o) $\dfrac{2}{5} \div \dfrac{6}{7} =$

p) $-\dfrac{5}{9} \div \dfrac{1}{3} =$

q) $-\dfrac{8}{9} \div -\dfrac{24}{27} =$

r) $\dfrac{3x}{4} \div \dfrac{6}{8} =$

s) $\dfrac{5}{6} \div \dfrac{10}{2y} =$

Lesson Four: Answers

a) $\dfrac{4}{12}$

b) $\dfrac{3}{9}$

c) b

d) $\dfrac{4}{6}, \dfrac{6}{9}, \dfrac{8}{12}$

e) d

f) $\dfrac{3}{7}$

g) $\dfrac{3}{4}$

h) $\dfrac{5}{6}$

i) $-\dfrac{5}{6}$

j) $-\frac{2}{3}$

k) $\frac{a}{b}$

l) $\frac{x}{2y}$

m) 4

n) 10

o) $\frac{1}{6}$

p) $\frac{1}{20}$

q) $\frac{1}{9}$

r) $\frac{6}{25}$

s) $-\frac{9}{28}$

t) $\frac{1}{6}$

u) $\frac{y}{42}$

v) $\frac{x}{25}$

w) $\frac{5}{6}$

x) $\frac{5}{4}$

y) $-\frac{10}{7}$

z) $\frac{1}{2}$

aa) $\frac{33}{64}$

bb) $\frac{7x}{15}$

cc) $\frac{15}{8y}$

dd) $\frac{y}{2}$

Lesson Four: Additional Exercises Answers

a) $\frac{6}{9}, \frac{9}{21}, \frac{12}{28}$

b) $\frac{2}{7}$

c) $\frac{2}{3}$

d) $\frac{7}{9}$

e) $-\frac{2}{5}$

f) $\frac{2}{3}$

g) $\frac{7x}{y}$

h) $\frac{1}{15}$

i) $\frac{6}{49}$

j) $\frac{2}{3}$

k) $-\frac{4}{15}$

l) $\frac{4x}{15}$

m) $\frac{y}{15}$

n) $\frac{3}{8}$

o) $\frac{7}{15}$

p) $-\frac{5}{3}$

q) 1

r) x

s) $\frac{y}{6}$

Lesson 5: Adding and Subtracting Fractions

As we discussed before, fractions are like slices of pie. This is $\frac{1}{4}$.

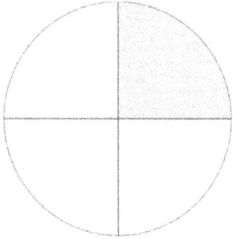

The bottom (denominator) is how the slices have been cut and the top (numerator) is how many slices we have. It wouldn't make sense to try to add slices that were different sizes. But if the slices are the same, you can.

$$\frac{1}{4} + \frac{2}{4}$$

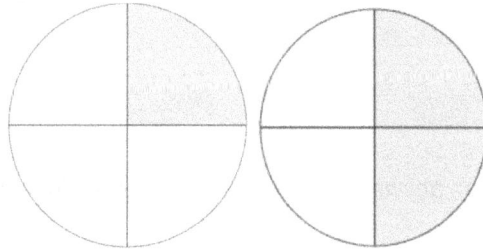

The size of the slices is the same. And, if we add them, we now have three. So:

$$\frac{1}{4} + \frac{2}{4} = \frac{3}{4}$$

Mathematically, it works like this. If the denominators are the same, we can simply add the numerators. Here's another.

$$\frac{2}{5} + \frac{1}{5} = \frac{3}{5}$$

Try some:

a) $\frac{3}{7} + \frac{2}{7} =$

b) $\frac{3}{12} + \frac{2}{12} =$

In this next problem, the new fraction can be reduced.

c) $\dfrac{1}{6}+\dfrac{2}{6}=$

Subtraction follows the exact same idea. If the denominators match, just subtract.

$$\frac{3}{5}-\frac{1}{5}=\frac{2}{5}$$

Try these:

d) $\dfrac{7}{11}-\dfrac{2}{11}=$

e) $\dfrac{15}{16}-\dfrac{2}{16}=$

On this problem, your answer can be reduced.

f) $\dfrac{8}{9}-\dfrac{2}{9}=$

This problem gets a negative for an answer. I know that is a strange idea when thinking of slices of pie, but mathematically it is okay.

g) $\dfrac{1}{4}-\dfrac{2}{4}=$

As always, we can extend the concepts to algebra.

$$\frac{x}{3}+\frac{1}{3}$$

It looks strange with an x, but it just means that I don't know how many slices I have in the first fraction. However, the denominators are the same, so I simply add.

$$\frac{x}{3}+\frac{1}{3}=\frac{x+1}{3}$$

But because the x and the 1 aren't like terms, this is the best I can do. Try some:

h) $\dfrac{y}{5}+\dfrac{3}{5}=$

i) $\dfrac{7}{15} + \dfrac{r}{15} =$

Subtraction works the same.

j) $\dfrac{x}{3} - \dfrac{1}{3} =$

k) $\dfrac{z}{7} - \dfrac{2}{7} =$

It is easy to add or subtract fractions when the denominators are the same. But what do we do when the denominators aren't the same?

$$\dfrac{1}{4} + \dfrac{1}{3}$$

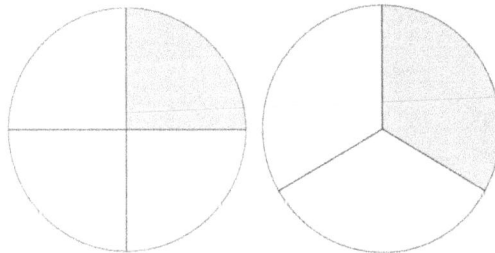

Adding two different sizes of pie would be meaningless. Instead, we need to re-slice the pieces so they match. In other words, we need to get a common denominator.

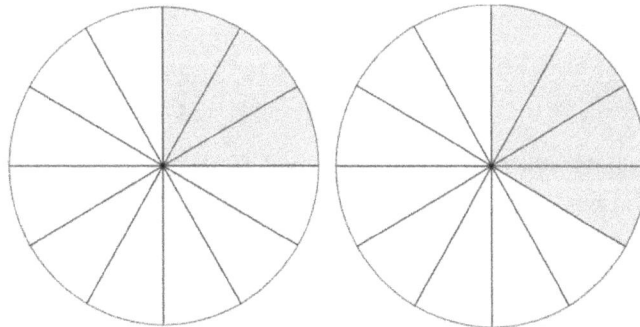

Now, with the slices the same, I can simply add.

$$\dfrac{3}{12} + \dfrac{4}{12} = \dfrac{7}{12}$$

But how did I determine the correct way to cut the pies? I used the same concept which we saw when we found Least Common Multiples. I found the lowest number that both denominators could multiply into. (Here it will be called finding the Least Common Denominator.) Here's a reminder of how we found the number.

Here are the prime factorizations of each of the original denominators.

$4: 2 \cdot 2$

$3: 3$

The first number they multiply into is the product of all their factors.

$$2 \cdot 2 \cdot 3 = 12$$

Multiplying all the factors would always give a common denominator, but not necessarily the least common denominator. Here is how we can be sure we find the LCD.

$$\frac{1}{6} + \frac{1}{8} =$$

Below are the prime factorizations.

$6: 2 \cdot 3$

$8: 2 \cdot 2 \cdot 2$

And, if they share any factors, circle them.

$6: \boxed{2} \cdot 3$

$8: \boxed{2} \cdot 2 \cdot 2$

Then, when you multiply your factors, any numbers in a circle only get counted once. So:

$2 \cdot 3 \cdot 2 \cdot 2 = 24$

Twenty-four is the lowest denominator that will work. It is the LCD- Least Common Denominator. Remember, you must make each fraction into an equivalent fraction. In other words, you can't just make the denominators 24 without changing the numerators.

$$\frac{1}{6} + \frac{1}{8} = \frac{1 \cdot 4}{6 \cdot 4} + \frac{1 \cdot 3}{8 \cdot 3} = \frac{4}{24} + \frac{3}{24} = \frac{7}{24}$$

Notice that for the denominator with a 6, I multiplied by 4. To the denominator with an 8, I multiplied by 3. Those are, of course, the numbers I need to multiply to get each to equal 24, but there is another reason. Look at the prime factorizations.

$6: \boxed{2} \cdot 3$

$8: \boxed{2} \cdot 2 \cdot 2$

1)   Why did the 6 need to be multiplied by 4 and the 8 need to be multiplied by 3?

Try some. (You can find the LCD in your head if you see it. However, it is very useful to learn how to do the prime factorization method. It will help us with harder problems as we move forward.)

m) $\dfrac{1}{6} + \dfrac{1}{14} =$

n) $\dfrac{3}{10} + \dfrac{4}{5} =$

o) $\dfrac{7}{8} + \dfrac{3}{10} =$

Subtraction works the same:

p) $\dfrac{1}{6} - \dfrac{1}{14} =$

q) $\dfrac{5}{12} - \dfrac{1}{8} =$

On this problem, you will end with a negative fraction. That is okay.

r) $\dfrac{1}{15} - \dfrac{3}{5} =$

Again, we can extend this to algebra.

$$\dfrac{x}{4} + \dfrac{1}{3}$$

The LCD is 12. And, even though it is difficult to wrap your head around, the process will be exactly the same.

$$\dfrac{x \cdot 3}{4 \cdot 3} + \dfrac{1 \cdot 4}{3 \cdot 4}$$

$$\dfrac{3x}{12} + \dfrac{4}{12} = \dfrac{3x + 4}{12}$$

I know the answer looks strange, but we followed all the same rules as adding normal fractions. (The $3x$ and the 4 are not like terms and so they couldn't be combined.)

Try some.

s) $\dfrac{y}{6} + \dfrac{1}{14} =$

t) $\dfrac{3}{10} + \dfrac{x}{5} =$

u) $\dfrac{z}{8} + \dfrac{3}{10} =$

Subtraction works the same:

v) $\dfrac{x}{6} - \dfrac{1}{14} =$

w) $\dfrac{5y}{12} - \dfrac{1}{8} =$

x) $\dfrac{1}{15} - \dfrac{3z}{5} =$

<u>Lesson Five: Additional Exercises:</u>

Add/Subtract the following fractions:

a) $\dfrac{1}{9} + \dfrac{4}{9} =$

b) $\dfrac{5}{12} + \dfrac{2}{12} =$

c) $\dfrac{3}{8} + \dfrac{1}{8} =$

d) $\dfrac{11}{15} - \dfrac{4}{15} =$

e) $\dfrac{7}{22} - \dfrac{3}{22} =$

f) $\dfrac{x}{5} + \dfrac{1}{5} =$

g) $\dfrac{3}{9} - \dfrac{y}{9} =$

h) $\dfrac{1}{3} + \dfrac{1}{6} =$

i) $\dfrac{3}{4} + \dfrac{1}{3} =$

j) $\dfrac{3}{14} + \dfrac{5}{21} =$

k) $\dfrac{1}{6} - \dfrac{1}{9} =$

l) $\dfrac{4}{5} - \dfrac{3}{4} =$

m) $\dfrac{11}{42} - \dfrac{1}{10} =$

n) $\dfrac{x}{14} + \dfrac{3}{6} =$

o) $\dfrac{3}{8} + \dfrac{y}{6} =$

p) $\dfrac{r}{15} - \dfrac{3}{10} =$

q) $\dfrac{5}{24} - \dfrac{s}{18} =$

Lesson Five: Answers

a) $\dfrac{5}{7}$

b) $\dfrac{5}{12}$

c) $\dfrac{1}{2}$

d) $\dfrac{5}{11}$

e) $\dfrac{13}{16}$

f) $\dfrac{2}{3}$

g) $-\dfrac{1}{4}$

h) $\dfrac{y+3}{5}$

i) $\dfrac{7+r}{15}$

j) $\dfrac{x-1}{3}$

k) $\dfrac{z-2}{7}$

l) The 6 would need the 2 • 2. The 8 would need the 3.

m) $\dfrac{10}{42} = \dfrac{5}{21}$

n) $\dfrac{11}{10}$

o) $\dfrac{47}{40}$

p) $\dfrac{2}{21}$

q) $\dfrac{7}{24}$

r) $-\dfrac{8}{15}$

s) $\dfrac{7y+3}{42}$

t) $\dfrac{3+2x}{10}$

u) $\dfrac{5z+12}{40}$

v) $\dfrac{7x-3}{42}$

w) $\dfrac{10y-3}{24}$

x) $\dfrac{1-9z}{15}$

Lesson Five: Additional Exercises Answers

a) $\dfrac{5}{9}$

b) $\dfrac{7}{12}$

c) $\dfrac{1}{2}$

d) $\dfrac{7}{15}$

e) $\dfrac{2}{11}$

f) $\dfrac{x+1}{5}$

g) $\frac{3-y}{9}$

h) $\frac{1}{2}$

i) $\frac{13}{12}$

j) $\frac{19}{42}$

k) $\frac{1}{18}$

l) $\frac{1}{20}$

m) $\frac{17}{105}$

n) $\frac{3x+21}{42}$

o) $\frac{9+4y}{24}$

p) $\frac{2r-9}{30}$

q) $\frac{15-4s}{72}$

a)   Add the following:

$$5 + 5 + 5 + 5 + 5 + 5 =$$

b)   Repeated addition is the same as multiplication. Which of the following would give the same answer as the problem above?

    a) $6 \cdot 5$

    b) $7 \cdot 5$

    c) $8 \cdot 5$

This works the same with variables.

$$x + x + x + x + x + x$$

c)   Which is the same?

    a) $5x$

    b) $7x$

    c) $6x$

Look at the following.

$$2 \cdot 5 + 6 \cdot 5$$

d)   Which is the same?

    a) $7 \cdot 5$

    b) $6 \cdot 5$

    c) $8 \cdot 5$

Look at this.

$$3x + 4x$$

e) Which is the same?

   a) $8x$

   b) $6x$

   c) $7x$

Subtraction doesn't change the idea. Simplify the following:

f) $12x - 4x + 6x =$

g) $-3x + 20x - 4x =$

One of the most foundational ideas in algebra is combining like terms. It simply means we can add matching variables. Only x's go with x's and y's with y's. Combine the following like terms:

h) $4x + 6y + 2x + 4y$

i) $14x - 3y + 12x - 2y$

The principal is the same even when things get more complicated. The key is to think of the variable portion as a word. To combine them, the word must be identical. Nothing combines in the following because even though they are similar, they are not the exact same word. One is x^2 and the other is x.

$$4x^2 + 3x$$

However, these have matches. Combine them.

$$4x^2 + 3x + 2x + 5x^2$$

j) Which is the correct answer?

   a) $7x^2 + 7x$

   b) $9x^2 + 5x$

   c) $8x^2 + 6x$

Combine these:

k) $14x^2 - 6x + 5x^2 + 3x - 2x^2$

l) $2x^2 + 7y + 5y^2 + 3x - 3y^2 + 6x^2$

m) $5xy + 6xy$

n) $11x^2y + 3x^2y$

o) $11x^2y + 3xy^2 - 2xy^2 + 10x^2y$

Be sure that you are matching identical "words." It is the key to combining like terms. Watch out for one more thing. If there is no number in front of a variable, it means that there is a one in front.

$$x + 5x = 6x$$

p) Combine like terms:

$$x^2 + y + 5y^2 + 3y + 2y^2 + 4x^2$$

Finally, look at this expression:

$$14x^2 + 2x + 3$$

There are three groups in this expression: $14x^2$, $2x$, 3. Groups are separated by + or -.

(Groups are numbers or variables being held together by multiplication or division.)

Coefficients are any numbers in front of variables. There is a coefficient of 14 in front of the x^2. And, there is a coefficient of 2 in front of the x. Although it doesn't have a variable, the 3 is still considered a coefficient.

List the terms and coefficients in this expression. (Hint: The 15 owns the negative sign, so it is -15. The x has a coefficient although it doesn't look like it. The coefficient is 1.)

$$6x^3 - 15x^2 + x + 9$$

q) Term:

r) Coefficients:

Lesson Six: Additional Exercises

For each of the following, combine like terms:

a) $3x + 4y + 12x + 7y$

b) $18z - 12t + 14z + 7 - 9t$

c) $x - 7y + 15y - 3x$

d) $3r - 8r + 15s - 9r + 3 - 7s + 21$

e) $16x^2 - 5x + 9y - 15x - 2x^2 + 11y^2 - 6y$

f) $-4m^2 + 15n^2 - 6m - 30n + 12m - 30n + 18m^2 - 3n^2$

g) $8xy + 7x - 3xy + 8x - 4y + 12xy + 4y$

h) $12x^2y - 25xy^2 + 30x^2y - 8x^2y + 5x^2 - 14y^2 - 6x^2 + 18x + 4xy^2$

Give the terms and coefficients for the following expressions:

i) $3x^3 - 5x^2 + x - 2$

j) $6y^4 + 9y^3 + 25y^2 - 2y + 105$

Lesson Six: Answers

a) 30

b) a

c) c

d) c

e) c

f) $14x$

g) $13x$

h) $6x + 10y$

i) $26x - 5y$

j) b

k) $17x^2 - 3x$

l) $8x^2 + 2y^2 + 7y + 3x$

m) $11xy$

n) $14x^2y$

o) $21x^2y + 1xy^2$

p) $5x^2 + 7y^2 + 4y$

q) $6x^3, -15x^2, x, 9$

r) $6, -15, 1, 9$

Lesson Six: Additional Exercises Answers

a) $15x + 11y$

b) $32z - 21t + 9$

c) $-2x + 8y$

d) $-14r + 8s + 24$

e) $14x^2 + 11y^2 - 20x + 3y$

f) $14m^2 + 12n^2 + 6m - 60n$

g) $15x + 17xy$

h) $-1x^2 + 34x^2y - 21xy^2 + 18x - 14y^2$

i) Terms: $3x^3, -5x^2, x, -2$

   Coefficients: $3, -5, 1, -2$

j) Terms: $6y^4, 9y^3, 25y^2, -2y, 105$

   Coefficients: $6, 9, 25, -2, 105$

Multiply the following:

a) $3 \cdot 5 =$

b) $3(2 + 3) =$

c) $(3 \cdot 2) + (3 \cdot 3) =$

It turns out that the following is true:

$$3(2 + 3) = (3 \cdot 2) + (3 \cdot 3)$$

Although it seems unnecessary, we could break a number into pieces and then multiply each piece.

$$3(2 + 3)$$

Here, I've multiplied the 3 to the first number inside and then to the second number inside. This is called the distributive property. It doesn't seem to have much of a point when multiplying two numbers, but in algebra it is very important.

$$5(x + 3)$$

$(x + 3)$ is a number, but I don't know what it is. The problem is the x. Here's where the distributive property comes in.

$$5(x + 3) = 5x + 15$$

Use the distributive property to simplify the following expressions:

d) $2(x + 5)$

e) $10(x - 2)$

f) $-3(x+4)$

g) $x(2+y)$

In these next two examples, we are actually multiplying a -1 through.

h) $-(3x+4)$

i) $-(2y-15)$

In these next problems, use the distributive property and then "clean-up" by combining any like terms.

j) $15+4(x-2)$

k) $-6-2(3x+7)$

l) $5x+3(6x-6)$

m) $-6y-6(2y+1)$

n) $3(2x-2)-6(5x+3)$

o) $12(2y+1)-(20y-8)$

Lesson Seven: Additional Exercises

Simplify using the distributive property:

a) $5(x - 7)$

b) $8(3 - y)$

c) $6(3x + 4)$

d) $3(4y + 2x)$

e) $-(2 + 8y)$

f) $-5(3 - 3x)$

g) $-4(x + y)$

h) $2y(3x - 5)$

i) $-12x - 2(3x + 5)$

j) $4(8y - 5) + 3(2 - y) - 2(5y - 1)$

Lesson Seven: Answers

a) 15

b) 15

c) 15

d) $2x + 10$

e) $10x - 20$

f) $-3x - 12$

g) $2x + xy$

h) $-3x - 4$

i) $-2y + 15$

j) $7 + 4x$

k) $-20 - 6x$

l) $23x - 18$

m) $-18y - 6$

n) $-24x - 24$

o) $4y + 20$

Lesson Seven: Additional Exercises Answers

a) $5x - 35$

b) $24 - 8y$

c) $18x + 24$

d) $12y + 6x$

e) $-2 - 8y$

f) $-15 + 15x$

g) $-4x - 4y$

h) $6xy - 10y$

i) $-18x - 10$

j) $19y - 12$

The ideas in this first section set a foundation for what is to come. Likely, it is a review for you. However, it is very important. The better you understand the terms and concepts; the easier the rest of algebra will be. This section will conclude the foundational work.

The most basic idea in algebra is the variable.

$$x + 3$$

We don't know the value of the number x. It could vary. Thus, the name variable. The 3 isn't varying, so it is called a constant.

In algebra, constants and variables will be combined to make one of two things: expressions or equations. Equations have an equal sign and can be solved to find a value for the variable. Expressions don't have an equal sign, and therefore, they can't be solved. The best we can do with an expression is to simplify it. This seems pretty easy, but I will often have students trying to solve expressions. It is a major problem and one to avoid.

a) Here, circle anything which is an equation. Put a square around anything which is an expression.

$$3x - 5$$

$$5y + 15 = 20$$

$$4z^2 - 7z + 18$$

$$\frac{r - 3}{r + 2} = 4$$

$$(x - 2)(x + 5) = 12$$

$$x^3 - 27$$

Exponents are repeated multiplication. For instance,

$$2^3 = 2 \cdot 2 \cdot 2 = 8$$

Give the value of each of the following exponents.

b) $3^4 =$

c) $12^2 =$

d) $5^3 =$

e) $(-2)^3 =$

Later, we will see problems involving inequalities. Variables can take on a range of values with an inequality.

$$x > 4$$

$$y \leq 2$$

Students often forget how the inequalities work. As in reading, math goes left to right. And, the larger opening is always toward the bigger number.

$$>$$

Greater Than

$$<$$

Less Than

$$\geq$$

Greater Than or Equal To

$$\leq$$

Less Than or Equal To

Math is a language, and learning it is like learning Spanish or Japanese. So, we often need to translate between math language and the English language. Below are a series of math expressions involving inequalities. Match each inequality on the left with the correct translation on the right.

f) $17 > 2$ x is less than or equal to two

g) $x \geq 8$ x is greater than or equal to eight

h) $-5 < 7$ seventeen is greater than two

i) $x \leq 2$ negative five is less than seven

Next, we will do some more translation from math language to English. We begin with addition and subtraction. Match the math expression on the left with its English translation on the right. (There are two right answers for each.)

Six more than x.

j) $x + 6$

Two subtracted from x.

k) $x - 2$ Six added to x.

Two less than x.

Translate each of the following math expressions into English:

l) $x + 15$

m) $y - 7$

69

Match some harder addition and subtraction translations: (There is only one correct answer per translation.)

n) $4x^2 + 15$ The difference of two x-squared and three y-squared.

o) $17x - 4y$ The difference of seventeen x and four y.

p) $9x^3 + 6$ The sum of four x-squared and fifteen.

q) $2x^2 - 3y^2$ The sum of nine x-cubed and six.

Translate each of the following math expressions into English:

r) $5x^2 + 21$

s) $5x^2 - 21$

We can also do multiplication and division. Match the following:

t) $8x$ The quotient of x-squared and six.

u) $\dfrac{8}{x}$ The product of eight and x.

v) $6x^2$ The quotient of eight and x.

w) $\dfrac{x^2}{6}$ The product of six and x-squared.

Translate each of the following math expressions into English:

x) $12x$

y) $\dfrac{x}{12}$

z) $\dfrac{12}{x}$

The hardest to translate are those involving grouping symbols. It will be a product times a sum. See if you can distinguish a difference as you match the following.

aa) $3(x + y)$ The sum of three times x and three times y.

bb) $3x + 3y$ Two times the difference of r and s.

cc) $2(r - s)$ The difference of two times r and two times s.

dd) $2r - 2s$ Three times the sum of x and y.

Translate each of the following math expressions into English:

ee) $12(a + b)$

ff) $12a - 12b$

Lesson Eight: Additional Exercises

Indicate if the following are expressions or equations:

a) $4x - 5 = 9$

b) $3x^2 = 63$

c) $6z^3 - 2z^2 + 14z - 1$

d) $y^2 + 4y - 5 = 0$

e) $x^2 - 9$

Give the value of the following exponents:

f) 2^5

g) $(-2)^5$

Translate each of the following from math language to English:

h) $r \leq 7$

i) $-3 > -5$

j) $x + 7$

k) $y - 3$

l) $3x^2 + 4$

m) $\dfrac{7}{b}$

n) $3(x - y)$

Lesson Eight: Answers

a) Equations: $5y + 15 = 20$; $\dfrac{r-2}{r+2} = 4$; $(x - 2)(x + 5) = 12$

b) 81

c) 144

d) 125

e) -8

f) Seventeen is greater than two

g) x is greater than or equal to 8

h) negative five is less than seven

i) x is less than or equal to two

j) six more than x; six added to x

k) two subtracted from x; two less than x

l) fifteen more than x; or x plus 15

m) seven less than y

n) The sum of four x-squared and fifteen

o) The difference of seventeen x and four y

p) The sum of nine x-cubed and six

q) The difference of two x-squared and three y-squared

r) The sum of five x-squared and twenty-one

s) The difference of five x-squared and twenty-one

t) The product of eight and x

u) The quotient of eight and x

v) The product of six and x-squared

w) The quotient of x-squared and six

x) The product of twelve and x

y) The quotient of x and twelve

z) The quotient of twelve and x

aa) Three times the sum of x and y

bb) The sum of three times x and three time y

cc) Two times the difference of r and s

dd) The difference of two times r and two times s

ee) Twelve times the sum of a and b; or the product of 12 and the sum of a and b.

ff) The difference of twelve times a and twelve times b

Lesson Eight: Additional Exercises Answers

a) equation

b) equation

c) expression

d) equation

e) expression

f) 32

g) -32

h) r is less than or equal to 7

i) negative 3 is greater than negative 5

j) x plus 7

k) 3 less than y

l) the sum of $3x^2$ and 4

m) 7 divided by b

n) 3 times the difference of x and y

Solving Equations

When students think of Algebra, this section is what comes to mind. There is a missing number, called a variable, and we want to find it. That's it. In a real sense, Algebra is a puzzle. If there is an equal sign involved, our goal is to discover the value which we don't know. In other words, solve for x. If you enroll in my College Algebra course, I need you to be completely confident in how to solve equations. By the end of this section, you will learn my three-step approach.

1) Clean the Rooms.
2) Get the x's back together.
3) Get the x alone.

For math strugglers, the main problem I encounter occurs in an equation like this one:

$$4x - 2 = 18$$

To get x alone, you must know what number needs to first leave the left side of the equation. Is it the 4? Or is it the 2? In my opinion, we typically teach this in the wrong order which prevents students from intuitively knowing which number leaves first. As such, we'll begin with an idea which I call a factor puzzle.

Solving Equations

Lesson 9: Factor Puzzles

As I've mentioned before, I believe the concept of factors is one of the key ideas in algebra. And factors can help us better understand the process for solving equations.

Here is what I call a factor puzzle:

$$2 \cdot x = 14$$

a) Two numbers are multiplying together to get the number 14. What is the missing number?

b) In algebra, we don't always show the multiplication sign when there is a variable. So, this next problem is multiplication again. Solve the factor puzzle. What value must y equal?

$$5y = 20$$

$$y =$$

Work the following. Find the missing value.

c) $8x = 72$

$x =$

d) $6z = 36$

$z =$

Try some that involve a negative.

e) $-2y = 18$

$y =$

f) $-7n = 42$

$n =$

g) $-3a = -18$

$a =$

If you aren't sure what the missing number is, all you need to do is divide.

$$11x = 132$$

$$x = \frac{132}{11} = 12$$

Try some that you may not know.

h) $24v = 312$

$v =$

i) $17x = 306$

$x =$

And if you had one which you didn't know involving a negative sign, just divide by the negative number.

j) $-16y = -224$

$y =$

k) $-19m = 76$

$m =$

Suppose your puzzle involved division. Some number divided by 7 equals 2.

$$\frac{x}{7} = 2$$

$$x = 14$$

Try these:

l) $\dfrac{y}{6} = 3$

$y =$

m) $\dfrac{z}{5} = 6$

$z =$

But if you ever are unsure of the missing number. Multiply.

$$\dfrac{y}{12} = 3$$

$$y = 3 \cdot 12 = 36$$

Work these.

n) $\dfrac{x}{15} = 5$

$x =$

o) $\dfrac{s}{16} = 13$

$s =$

Try some that involve negative numbers. Again, if you can't see them in your head, just multiply.

p) $\dfrac{y}{-4} = 8$

$y =$

q) $\dfrac{x}{-22} = -11$

$x =$

Look at this factor puzzle.

$$-m = 15$$

This is really the same as:

$$-1m = 15$$

In this case, the value of m must be:

$$m = -15$$

r) Try some. Remember, if you aren't sure, just divide by -1.

$$-x = 21$$

$$x =$$

s) $$-r = -7$$

$$r =$$

The next group of problems aren't technically factor puzzles. They involve fractions, and fractions don't divide evenly, which factors must do. However, the concept for solving the missing number remains the same. Just divide.

$$\frac{2}{3}x = \frac{1}{4}$$

So, we are dividing fractions:

$$x = \frac{\frac{1}{4}}{\frac{2}{3}}$$

And when we divide fractions, we Keep, Change, Flip.

$$x = \frac{1}{4} \cdot \frac{3}{2} = \frac{3}{8}$$

Work these. Remember, if your answer is a fraction that can be reduced, do so:

t) $$\frac{3}{5}y = \frac{2}{3}$$

$$y =$$

u) $\dfrac{4}{3}n = \dfrac{2}{9}$

$n =$

v) $-\dfrac{1}{2}x = \dfrac{1}{2}$

$x =$

w) $-\dfrac{7}{2}s = -\dfrac{14}{5}$

$s =$

Another key idea in solving equations is to "clean up." We will explore this idea more later, but for now, it means to combine like terms. If you need a refresher on combining like terms, return to that activity. Look at this problem.

$$5x + 2x = 21$$

There are terms on the left which are alike and can be combine.

$$7x = 21$$

After we cleaned up the like terms, we have a factor puzzle.

$$x = 3$$

Try some:

x) $5y - 2y = 18$

y) $12m + 2m - 4m = 100$

This problem requires combining like terms on both sides.

z) $-3x + 5x = 8 + 6$

aa) $3y - 8y + 10y = 20 - 25$

On these next problems, the variable is on the right instead of the left. Nothing changes. The puzzle remains the same.

bb) $-121 = 14s - 3s$

cc) $15 - 4 + 9 = 18y - 16y + 2y$

<u>Lesson Nine: Additional Exercises</u>

a) $3x = 21$

b) $5y = 45$

c) $2s = -14$

d) $-7r = 28$

e) $-4x = -32$

f) $15v = 165$

g) $-8k = 128$

h) $\dfrac{x}{4} = 7$

i) $\dfrac{y}{9} = -8$

j) $\dfrac{r}{-6} = -12$

k) $\dfrac{x}{-18} = -14$

l) $\dfrac{2}{3}x = \dfrac{4}{5}$

m) $-\dfrac{5}{7}y = -\dfrac{1}{14}$

n) $8x - 6x = 14$

o) $3y - 14y + 7y = -64$

p) $-4r + 17r - 6r = 57 - 8$

q) $-21 = 15x - 20x + 2x$

Lesson Nine: Answers

a) 7

b) $y = 4$

c) $x = 9$

d) $z = 6$

e) $y = -9$

f) $n = -6$

g) $a = 6$

h) $v = 13$

i) $x = 18$

j) $y = 14$

k) $m = -4$

l) $y = 18$

m) $z = 30$

n) $x = 75$

o) $s = 208$

p) $y = -32$

q) $x = 242$

r) $x = -21$

s) $r = 7$

t) $y = \dfrac{10}{9}$

u) $n = \dfrac{1}{6}$

v) $x = -1$

w) $s = \dfrac{4}{5}$

x) $y = 6$

y) $m = 10$

z) $x = 7$

aa) $y = -1$

bb) $s = -11$

cc) $y = 5$

Lesson Nine: Additional Exercises Answers

a) $x - 7$

b) $y = 9$

c) $s = -7$

d) $r = -4$

e) $x = 8$

f) $v = 11$

g) $k = -16$

h) $x = 28$

i) $y = -72$

j) $r = 72$

k) $x = 252$

l) $x = \dfrac{6}{5}$

m) $y = \dfrac{1}{10}$

n) $x = 7$

o) $y = 16$

p) $r = 7$

q) $x = 7$

As we saw in our last activity, factors are two numbers that multiply to create another number. Here is an example:

$$3 \cdot 5 = 15$$

And, we looked at an idea which I called factor puzzles. If we had two numbers multiplied together and we didn't know one, we would have a factor puzzle.

$$3 \cdot x = 15$$

I teach it this way, because it is fairly easy to see that x must be 5. For this activity, we are going to see equations like this:

$$x + 3 = 7$$

The x and 3 are not factors of 7. The addition prevents that. Factors must be two numbers which

a) multiply together to make the number. Circle any of the following which involve factors:

$$3 \cdot y = 21$$

$$z - 2 = 18$$

$$5 + k = 3$$

$$9x = 81$$

Here we don't have factors, but we still want to discover what the missing number is.

$$x + 3 = 7$$

You can probably do this in your head. But, if you couldn't, you simply take away the 3. However, the equal sign means this is like a balance:

$$x \; + \; 3 \quad = \quad 7$$

If you remove 3 from one side, it won't remain balanced unless you remove (subtract) 3 from the other side.

$$x \quad = \quad 4$$

Solve these:

b) $y + 7 = 14$

c) $r + 15 = 30$

d) $c + 6 = 18$

And, in the same manner, it can go the other way.

$$x - 5 = 10$$

To discover x, we add 5. But if we do it to one side, we must do it to the other in order to keep things balanced.

$$x - 5 + 5 = 10 + 5$$

$$x = 15$$

Solve these:

e) $y - 2 = 8$

f) $s - 12 = 20$

g) $a - 7 = 21$

This next set involves negative numbers, but the concept is the same.

$$x + 5 = -2$$
$$x = -2 - 5$$
$$x = -7$$

Work these:

h) $y + 8 = -10$

i) $t - 7 = -12$

j) $b + 3 = -21$

The idea carries over to fractions and decimals too.

$$x + \frac{1}{2} = \frac{3}{4}$$

$$x = \frac{3}{4} - \frac{1}{2}$$

$$x = \frac{3}{4} - \frac{2}{4} = \frac{1}{4}$$

Solve the following:

k) $y + \dfrac{2}{3} = \dfrac{5}{6}$

l) $r - \dfrac{1}{4} = \dfrac{3}{8}$

m) $c + 2.1 = 9.3$

n) $d - 3.4 = 7.1$

o) $x + 4.5 = -2.6$

As we will also discuss more as we work with harder equations, the next step is to "clean-up." This means to combine like terms.

$$5x - 4x = 9$$

$$x = 9$$

They can get more complicated:

$$9x + 5 - 8x - 2 = 12$$

Clean up. (Combine like terms.)

$$x + 3 = 12$$

$$x = 9$$

Try these:

p) $7y - 6y = 14$

q) $8z + 5 - 7z = 9$

r) $9x - 5 - 8x = -12$

s) $7s + 12 - 7 + s - 7s = 2$

t) $2x - 14 + 7 - x = 12 - 6$

This last set of problems requires the distributive property. Clean-up by multiplying and then combining like terms.

$$3(x - 3) - 2x = 7$$
$$3x - 9 - 2x = 7$$
$$x - 9 = 7$$
$$x = 16$$

Work these:

u) $4(2x + 2) - 7x = 15$

v) $3(y + 5) - 2y + 7 = -12$

w) $5(v - 2) + 17 - 4v = 12 + 3$

On this problem, remember that the 2 owns the negative sign in front of it. It acts like a -2 as it gets distributed.

x) $3(x + 7) - 2(x - 5) = 3$

With this final problem, clean up both rooms. The variable will disappear from the right side.

y) $4(y - 2) - 3(y + 1) = 5(y + 1) - 5(y + 2)$

Lesson Ten: Additional Exercises

Solve.

a) $x + 5 = 55$

b) $y - 7 = 16$

c) $m - 8 = -21$

d) $s + 7 = -35$

e) $x - \frac{1}{4} = \frac{2}{3}$

f) $y + .8 = -1.3$

g) $15x - 14x + 8 - 3 = 12$

h) $7y - 3 - 6y - 9 = 20$

i) $-2w + 11 + 3w - 15 = 8 - 17$

j) $-12 + 23x - 8 + 5x - 2 - 27x = -19 - 2$

k) $3(x - 2) - 2x = 7$

l) $-2(y + 5) + 13 + 3y - 7 = -1$

m) $3(r + 5) - 2(r - 6) = 22$

n) $-(12v - 11) - (-13v + 12) = 1$

<u>Lesson Ten: Answers</u>

a) $3y = 21; 9x = 81$

b) $y = 7$

c) $r = 15$

d) $c = 12$

e) $y = 10$

f) $s = 32$

g) $a = 28$

h) $y = -18$

i) $t = -5$

j) $b = -24$

k) $y = \frac{1}{6}$

l) $r = \frac{5}{8}$

m) $c = 7.2$

n) $d = 10.5$

o) $x = -7.1$

p) $y = 14$

q) $z = 4$

r) $x = -7$

s) $s = -3$

t) $x = 13$

u) $x = 7$

v) $y = -34$

w) $v = 8$

x) $x = -28$

y) $y = 6$

Lesson Ten: Additional Exercises Answers

a) $x = 50$

b) $y = 23$

c) $m = -13$

d) $s = -42$

e) $x = \frac{11}{12}$

f) $y = -2.1$

g) $x = 7$

h) $y = 32$

i) $w = -5$

j) $x = 1$

k) $x = 13$

l) $y = 3$

m) $r = -5$

n) $v = 2$

Lesson 11: Equations with Variables on Both Sides

Let's continue to solve equations. Our goal is to get this equation down to a factor puzzle, but the 5 is preventing it. But, as we saw in the last activity, we know how to get rid of the 5.

$$2x + 5 = 15$$

Subtract the 5 from both sides.

$$2x = 15 - 5$$

$$2x = 10$$

Now we are left with a factor puzzle which we can easily solve.

$$x = 5$$

Solve. (First, get the problem down to a factor puzzle.)

a) $3y - 7 = 14$

b) $5z + 3 = 28$

c) $6r - 2 = 34$

d) $8x - 5 = -69$

e) $-2y + 4 = -16$

Look at our next problem.

$$5x = 3x + 14$$

We don't have a factor puzzle yet. To make one, we need to get all the x terms back together. In order to do that, we must remove $3x$ from the right. But if we do something to the right, we must do the same thing to the left.

$$5x - 3x = 3x - 3x + 14$$

$$2x = 14$$

Now we have a factor puzzle which we can solve.

$$x = 7$$

Work these:

f) $7y = 3y + 24$

g) $6r = 2r - 12$

h) $8x = x + 49$

In these next problems, to get the variables back together, we need to add instead of subtract.

$$4y = -2y + 18$$

$$4y + 2y = -2y + 2y + 18$$

$$6y = 18$$

$$y = 3$$

Solve the following:

i) $7r = -3r + 40$

j) $4z = -2z - 36$

k) $-3x = -2x - 7$

Finally, it isn't necessary to always move the variable to the room on the left. We may need to go the other direction. The process works the same.

$$5y + 8 = 3y$$

If we moved the $3y$ to the right, it would leave that room empty. Instead, we move the $5y$.

$$5y - 5y + 8 = 3y - 5y$$

$$8 = 2y$$

Our factor puzzle is reversed but the concept remains.

$$y = 4$$

Work these:

l) $2x + 4 = 6x$

m) $3c - 12 = 5c$

n) $-7w - 20 = 3w$

o) $-9x + 17 = -10x$

Lesson Eleven: Additional Exercises

a) $3x + 5 = 23$

b) $7y - 12 = 37$

c) $8r - 6 = -46$

d) $-5x - 7 = 28$

e) $11m = 7m + 16$

f) $-2s = -5s + 27$

g) $-x + 30 = 9x$

h) $8x - 16 = 4x$

i) $16y + 64 = 20y$

j) $-3p - 42 = -6p$

Lesson Eleven: Answers

a) $y = 7$

b) $z = 5$

c) $r = 6$

d) $x = -8$

e) $y = 10$

f) $y = 6$

g) $r = -3$

h) $x = 7$

i) $r = 4$

j) $z = -6$

k) $x = 7$

l) $x = 1$

m) $c = -6$

n) $w = -2$

o) $x = -17$

Lesson Eleven: Additional Exercises Answers

a) $x = 6$

b) $y = 7$

c) $r = -5$

d) $x = -7$

e) $m = 4$

f) $s = 9$

g) $x = 3$

h) $x = 4$

i) $y = 16$

j) $p = 14$

Lesson 12: Moving both Variables and Constants

The next step in solving equations involves moving both variables and constants to the other side.

$$7x + 8 = 5x + 12$$

First, we need to get the x's back together.

$$7x - 5x + 8 = 5x - 5x + 12$$

$$2x + 8 = 12$$

Now we need to get down to a factor puzzle.

$$2x + 8 - 8 = 12 - 8$$

$$2x = 4$$

And solving the factor puzzle is easy.

$$x = 2$$

Work these:

a) $5y + 2 = 3y + 16$

b) $7z - 6 = 2z + 19$

c) $11v - 7 = 2v - 25$

d) $-2x + 1 = -8x + 43$

I believe the idea of a factor puzzle helps us to understand the order in which we are to solve equations. However, as I've mentioned before, the factor puzzle concept doesn't always involve true factors.

$$3y = 7$$

$$y = \frac{7}{3}$$

Factors are two number that divide evenly, yet we got a fraction for y. Despite this limitation, use the factor puzzle concept to help you work these problems. You will get fractions.

e) $5x + 13 = -2x - 7$

f) $-12r + 4 = -8r - 15$

And, the ideas can extend to fractions and decimals. Use the same process to solve these.

g) $\frac{1}{2}x + 11 = \frac{1}{4}x + 19$

h) $\frac{2}{3}y - 2 = -\frac{1}{6}y + 4$

i) $2.1x + 3 = 1.5x + 9$

j) $1.7y - 5 = 1.8y + 11$

Lesson Twelve: Additional Exercises

Solve the following equations:

a) $7x + 20 = 3x - 4$

b) $3y - 21 = 5y - 13$

c) $-9r - 51 = -2r + 5$

d) $-6s + 35 = 7s - 4$

e) $8x + 9 = -3x + 20$

f) $11p + 27 = 3 - p$

g) $21y - 18 = 2 + 5y$

h) $-19v - 45 = 2v + 39$

i) $\frac{5}{6}x + 2 = -\frac{1}{12}x - 9$

j) $.4y - 1.8 = .8y - .6$

Lesson Twelve: Answers

a) $y = 7$

b) $z = 5$

c) $v = -2$

d) $x = 7$

e) $x = -\dfrac{20}{7}$

f) $r = \dfrac{19}{4}$

g) $x = 32$

h) $y = \dfrac{36}{5}$

i) $x = 10$

j) $y = -160$

Lesson Twelve: Additional Exercises Answers

a) $x = -6$

b) $y = -4$

c) $r = -8$

d) $s = 3$

e) $x = 1$

f) $p = -2$

g) $y = \dfrac{5}{4}$

h) $v = -4$

i) $x = -12$

j) $y = -3$

Solving Equations

Lesson 13: Solving Complicated Equations

As equations become more complicated, the process to solve them is always the same. Follow these three steps for any equation:

1) Clean Up Both Rooms- Do any distributing and combining like terms on both sides of the equation.
2) Get the Variable Back Together- If there are variables on both sides of the equation, move them all to one side.
3) Get the Variable Alone- Get down to a factor puzzle and solve.

Here are a couple of problems which show the steps.

$$3(y - 2) + 7 = y - 9$$

1) Clean Up Both Rooms.

$$3y - 6 + 7 = y - 9$$

$$3y + 1 = y - 9$$

The rooms are now clean. Next, there are variables in both rooms.

2) Get the Variable Back Together.

$$3y - y + 1 = -9$$

$$2y + 1 = -9$$

3) Get the Variable Alone. Make a factor puzzle and then solve.

$$2y = -10$$

$$y = -5$$

Here is another.

$$8(x + 3) + 4 = 2(x - 2) + 2$$

1) Clean Up Both Rooms.

$$8x + 24 + 4 = 2x - 4 + 2$$

$$8x + 28 = 2x - 2$$

2) Get the Variable Back Together.

$$8x - 2x + 28 = -2$$

$$6x + 28 = -2$$

3) Get the Variable Alone. Make a factor puzzle and then solve.

$$6x = -2 - 28$$

$$6x = -30$$

$$x = -5$$

Now, solve some on your own. If you don't need a step, just skip it. In this first problem, after cleaning up, there aren't variables in both rooms. Just move on to Get the Variable Alone.

a) $4(r + 5) + 4 = 36$

b) $2(2x - 3) + 7 = 2(x + 5) - 1$

c) $10 - 4(x - 3) = 14x - 14$

d) $2(2y - 1) + 2 = 5(y - 7) + 5$

In this next problem, the solution will be a fraction. However, the process remains the same.

e) $2r + 3(r - 1) + 7 = 2(r + 5) - 8$

This final problem will involve parenthesis inside of parenthesis. Clean up from the inside of the room outward.

f) $2[3 - 2(x - 4)] = 6x + 2$

<u>Lesson Thirteen: Additional Exercises</u>

Solve the following:

a) $2(x - 3) + 5 = 3(x + 4)$

b) $12 + 5(y + 3) + 2 = -2(y + 10)$

c) $-(2x - 7) - 3(2x + 7) = 4x + 1 + 3x$

d) $7y + 4(3y - 4) - 3y = 64$

e) $-2(3x - 4) + 5(2x + 6) = 3(x + 1) + 21$

f) $7(2v + 2) = 4(3v - 5) - 3(v - 3)$

g) $12 + 8(-3r - 4) = -10(r + 4)$

h) $18 - 3(3y - 11) + 7 = 4(2y + 11)$

i) $3[4 - 2(3x - 4)] = -3(4 + 2x)$

j) $5[2 + 3(4y - 5)] - 35 = 25(2y + 3) + 25$

a) $r = 3$

b) $x = 4$

c) $x = 2$

d) $y = 30$

e) $r = -\dfrac{2}{3}$

f) $x = 2$

Lesson Thirteen: Additional Exercises Answers

a) $x = -13$

b) $y = -7$

c) $x = -1$

d) $y = 5$

e) $x = -14$

f) $v = -5$

g) $r = \dfrac{10}{7}$

h) $y = \dfrac{14}{17}$

i) $x = 4$

j) $y = 20$

Factoring

In Algebra, sometimes we are trying to discover the value of a missing number. However, if we are working with an expression, without an equal sign, we can't solve. But that doesn't mean there is no math that can be done. First, we saw that Algebra is *solving* for missing numbers. Now, we will see that Algebra is also *working* with missing numbers. We've already seen this with the distributive property.

$$3(x + 2) = 3x + 6$$

Here, $(x + 2)$ is a number which we don't know. We don't know its value because it includes the variable x. But, as the distributive property says, it turns out that multiplying 3 by each "piece" of $x + 2$ is valid.

All of the ideas you learned in arithmetic can still be done even when a number (or part of a number) is unknown. You can add, subtract, multiply, and divide. But, perhaps the most important thing you can do is factor.

$$18 = 2 \cdot 3 \cdot 3$$

$$x^2 - 9 = (x - 3)(x + 3)$$

Because it includes x, we don't know the value of the number $x^2 - 9$, but this mystery number is still composite and has two factors: $(x - 3)$ and $(x + 3)$. Before we continue, here's an important idea that will help you a great deal.

Just as we saw when trying to solve equations, addition and subtraction are a difficulty. They are preventing us from having a simple factor puzzle. Factors consist only of multiplication. But addition and subtraction join. For instance, $3x$ is multiplication, so it is easy to factor.

$$3x = 3 \cdot x$$

But the subtraction in $x^2 - 9$ has joined the x^2 and the 9. They are now combined into one new number and factoring isn't so simple. Here is the important take away. Each of the following are only multiplication, so they can be easily factored:

$$5y = 5 \cdot y$$

$$7xy = 7 \cdot x \cdot y$$

$$9a^2b^3 = 3 \cdot 3 \cdot a \cdot a \cdot b \cdot b \cdot b$$

Yet, each of these next numbers (called polynomials) have been joined together with addition or subtraction, which will require us to learn some advanced strategies in order to factor.

$$x^2 - x - 6 = (x - 3)(x + 2)$$

$$a^2 - 121 = (a - 11)(a + 11)$$

$$y^3 + 27 = (y + 3)(y^2 + 3y + 9)$$

So, watch out for addition and subtraction. They combine numbers and make the work of factoring more challenging.

I named this section factoring because I believe it is one of the most important aspects of Algebra. However, if factoring is breaking a number down into multiplication, understanding how we multiply in Algebra seems like an important place to start.

Lesson 14: Multiplying with a Monomial

Look at the following.

$$2 \cdot 2 \cdot 2 \cdot 2 = 2^4$$

When we multiply variables, we are doing the same thing.

a) $x \cdot x \cdot x \cdot x = x^4$

If there are exponents, we could imagine them written out like this. (Finish this problem.)

b) $y^2 \cdot y^3 = y \cdot y \cdot y \cdot y \cdot y =$

So, when multiplying a variable, we simply need to add the exponents. Work this problem.

c) $a^5 \cdot a^7 \cdot a =$

And, if there are numbers in front of the variable (coefficients), since the order we multiply doesn't matter, we first multiply the numbers and then combine the variables.

$$2x^2 \cdot 5x^3 - 2 \cdot 5 \cdot x^2 \cdot x^3 - 10x^5$$

Try these:

d) $6x^3 \cdot 4x^5$

e) $10y^5 \cdot 5y^3$

f) $2x \cdot 7x^8$

g) $4y^2 \cdot 6y^3 \cdot y$

h) $-3x \cdot 8x^2$

i) $-5x \cdot -3x$

We can extend the idea by multiplying a monomial by a polynomial. First, a quick refresher on the distributive property. Simplify the following:

j) $3(x - 5)$

k) $2(5x^2 + 6x - 2)$

l) $6(4xy + 6)$

When we multiply a monomial (one term) by a polynomial (multiple terms), we use the distributive property.

$$4x(3x + 2) = (4x)(3x) + (4x)(2) = 12x^2 + 8x$$

Simplify these:

m) $5x(10x - 3)$

n) $2y^2(4y + 7)$

o) $3a(8a^2 - 2a + 5)$

p) $4xy(9xy - 2)$

q) $7x^2y(2x^2y + x - 1)$

r) $-2x^2y(x^2 + 5xy + y^2)$

Lesson Fourteen: Additional Exercises

Multiply the following:

a) $a^7 \cdot a^6 \cdot a^2$

b) $12x \cdot 3x \cdot 2x$

c) $-11v^2 \cdot 5v^4$

d) $5xy \cdot 6x^2$

e) $-2r^4s^3 \cdot 11r^7s^5$

f) $6(x+4)$

g) $-9(3xy+5)$

h) $8x(3x-2)$

i) $-5xy(11x^2y^2-3xy)$

j) $2xy^2(15x^2-7x+5)$

k) $15r^3(r^2-r-1)$

l) $-9x^2y^3z^5(4x^2y^2z-3x^2yz+6xyz-2z^2)$

Lesson Fourteen: Answers

a) x^4

b) y^5

c) a^{13}

d) $24x^8$

e) $50y^8$

f) $14x^9$

g) $24y^6$

h) $-24x^3$

i) $15x^2$

j) $3x - 15$

k) $10x^2 + 12x - 4$

l) $24xy + 36$

m) $50x^2 - 15x$

n) $8y^3 + 14y^2$

o) $24a^3 - 6a^2 + 15a$

p) $36x^2y^2 - 8xy$

q) $14x^4y^2 + 7x^3y - 7x^2y$

r) $-2x^4y - 10x^3y^2 - 2x^2y^3$

Lesson Fourteen: Additional Exercises Answers

a) a^{15}

b) $72x^3$

c) $-55v^6$

d) $30x^3y$

e) $-22r^{11}s^8$

f) $6x + 24$

g) $-27xy - 45$

h) $24x^2 - 16x$

i) $-55x^3y^3 + 15x^2y^2$

j) $30x^3y^2 - 14x^2y^2 + 10xy^2$

k) $15r^5 - 15r^4 - 15r^3$

l) $-36x^4y^5z^6 + 27x^4y^4z^6 - 54x^3y^4z^6 + 18x^2y^3z^7$

Multiply the following:

a)
$$12x(x - 5)$$

b)
$$4x^2(x^2 - 2x + 3)$$

To multiply a binomial by a binomial, the idea behind these problems is the distributive property. Simplify the following:

c)
$$x(x + 3)$$

d)
$$2(x + 3)$$

So, to multiply a binomial by a binomial, we are just going to do the distributive property twice.

$$(x + 2)(x + 3) = x(x + 3) + 2(x + 3)$$

Finish this by multiplying and combining like term:

e)
$$x(x + 3) + 2(x + 3) =$$

Try these. Do the distributive property twice:

f) $(2x + 5)(3x + 2)$

g) $(5y - 2)(10y + 4)$

h) $(7a + 3)(a^2 - 1)$

i) $(2ab - 1)(3ab + 1)$

j) $(x - 3)(x + 3)$

And, of course, the idea can be extended further.

$$(x + 5)(2x^2 + 4x - 2)$$

You can distribute three times. Here it is in parts. Simplify each part.

k) $x(2x^2 + 4x - 2)$

l) $5(2x^2 + 4x - 2)$

m)   Now, combine the like terms from each of your two answers.

Try these:

n) $(x + 3)(x^2 + 5x + 1)$ o) $(x - 1)(x^2 + 3x - 1)$

Lesson Fifteen: Additional Exercises

Simplify.

a) $(3x + 7)(2x + 5)$

b) $(4y - 7)(5y + 2)$

c) $(z - 1)(z - 1)$

d) $(8p + 9)(8p - 9)$

e) $(7a - 3)(7a - 3)$

f) $(2x^2 - 6)(x + 3)$

g) $(p^2 - 1)(p^2 + 1)$

h) $(5xy + 1)(3xy + 2)$

i) $(y - 1)(y^2 + y + 1)$

j) $(2s - 2)(3s^2 - 4s + 1)$

Lesson Fifteen: Answers

a) $12x^2 - 60x$

b) $4x^4 - 8x^3 + 12x^2$

c) $x^2 + 3x$

d) $2x + 6$

e) $x^2 + 5x + 6$

f) $6x^2 + 19x + 10$

g) $50y^2 - 8$

h) $7a^3 + 3a^2 - 7a - 3$

i) $6a^2b^2 - ab - 1$

j) $x^2 - 9$

k) $2x^3 + 4x^2 - 2x$

l) $10x^2 + 20x - 10$

m) $2x^3 + 14x^2 + 18x - 10$

n) $x^3 + 8x^2 + 16x + 3$

o) $x^3 + 2x^2 - 4x + 1$

Lesson Fifteen: Additional Exercises Answers

a) $6x^2 + 29x + 35$

b) $20y^2 - 27y - 14$

c) $z^2 - 2z + 1$

d) $64p^2 - 81$

e) $49a^2 - 6a + 9$

f) $2x^3 + 6x^2 - 6x - 18$

g) $p^4 - 1$

h) $15x^2y^2 + 13xy + 2$

i) $y^3 - 1$

j) $6s^3 - 14s^2 + 10s + 23$

Factoring

Lesson 16: Multiplying Two of the Same Binomial

In this lesson, we will look at a special outcome which happens when we multiply some binomials. It isn't necessary to memorize it. You begin to recognize it simply from repetition. Multiply the following:

a) $(x + 3)(x + 3) =$

b) $(y - 2)(y - 2) =$

c) $(2x + 5)(2x + 5) =$

d) What is true about the two binomials we are multiplying?

With these binomials, you always create a trinomial. And, here's the pattern.

1) Square of the first term.
2) Square of the last term.
3) Two of the first term times the last term.

This is true, but, again, you don't need to memorize it. Just multiply it out. You will always get the same result. Try a few more:

e) $(2x - 3)(2x - 3)$

f) $(x^2 - 1)(x^2 - 1)$

g) $(x - 6)^2$

h) $(2x - 3y)(2x - 3y)$

i) $(2m^2 + 8)^2$

Lesson Sixteen: Additional Exercises

Simplify.

a) $(x+5)(x+5)$

b) $(y-6)(y-6)$

c) $(z+9)(z+9)$

d) $(2x-1)(2x-1)$

e) $(x+2)^2$

f) $(y^2-3)(y^2-3)$

g) $(3x-6y)(3x-6y)$

h) $(4r^2-3s^2)^2$

Lesson Sixteen: Answers

a) x^2+6x+9

b) y^2-4y+4

c) $4x^2+20x+25$

d) They are the same.

e) $4x^2-12x+9$

f) $x^4 - 2x^2 + 1$

g) $x^2 - 12x + 36$

h) $4x^2 - 12xy + 9y^2$

i) $4m^4 + 32m^2 + 64$

Lesson Sixteen: Additional Exercises Answers

a) $x^2 + 10x + 25$

b) $y^2 - 12y + 36$

c) $z^2 + 18z + 81$

d) $4x^2 - 4x + 1$

e) $x^2 + 4x + 4$

f) $y^4 - 6y^2 + 9$

g) $9x^2 - 36xy + 36y^2$

h) $16r^4 - 24r^2s^2 + 9s^4$

Lesson 17: Multiplying Conjugate Binomials

In the last activity, we saw a special product which occurs whenever we multiply two of the same binomials. In this activity, we are going to find another special product. Again, it doesn't need to be memorized, although after doing enough of them, you will begin to know the pattern by heart.

Multiply the following:

a) $(x + 3)(x - 3) =$

b) $(y - 2)(y + 2) =$

c) $(2x + 5)(2x - 5) =$

d)   What is true about the two binomials we are multiplying?

e)   After you have multiplied them, what is always true of the sign in the middle?

In this special case, you always create a trinomial. And, here's the pattern.

1) Square of the first term.
2) Square of the last term.
3) Two opposite middle terms which cancel out.
4) In the final answer, the sign between the first and last term is always negative.

Again, this is true, but you don't need to memorize it. Just multiply it out. You will always get the same result. Try a couple more:

f) $(2x - 3)(2x + 3)$

g) $(x^2 + 1)(x^2 - 1)$

h) $(2x - 3y)(2x + 3y)$

i) $(2m^2 + 8)(2m^2 - 8)$

What we are multiplying here are called conjugates. The two terms are the same but they have opposite signs in-between. Conjugates are useful because when conjugates are multiplied the middle term will always drop out. Give the conjugate for each of the following:

j) $(x + 9)$

k) $(2y - 7)$

l) $(a^2 + b^2)$

Multiply the following:

a) $(x-4)(x+4)$

b) $(y+7)(y-7)$

c) $(2v-5)(2v+5)$

d) $(7r+11)(7r-11)$

e) $(x-y)(x+y)$

f) $(4x-3y)(4+3y)$

g) $(x^2-9)(x^2+9)$

h) $(3x^2-8y^2)(3x^2+8y^2)$

Lesson Seventeen: Answers

a) $x^2 - 9$

b) $y^2 - 4$

c) $4x^2 - 25$

d) same but opposite signs in the middle

e) always negative

f) $4x^2 - 9$

g) $x^4 - 1$

h) $4x^2 - 9y^2$

i) $4m^4 - 64$

j) $(x - 9)$

k) $(2y + 7)$

l) $a^2 - b^2$

Lesson Seventeen: Answers

a) $x^2 - 16$

b) $y^2 - 49$

c) $4x^2 - 25$

d) $49r^2 - 121$

e) $x^2 - y^2$

f) $16x^2 - 9y^2$

g) $x^4 - 81$

h) $9x^4 - 64y^4$

Factoring

Lesson 18: Greatest Common Factor

a)   What is the largest number which you could divide into 12 and 15?

Below, I have created the prime factorization for the numbers 12 and 15. Draw a circle around any shared factors.

$$12 = 2 \cdot 2 \cdot 3$$

$$15 = 5 \cdot 3$$

This shared factor is called the **GCF** (Greatest Common Factor).

b)   What is the largest number you could divide into 12 and 30?

Once again, I have created the prime factorization for the numbers. Draw a circle around any shared factors.

$$12 = 2 \cdot 2 \cdot 3$$

$$30 = 2 \cdot 3 \cdot 5$$

If you multiply the shared factors together and you have the GCF.

Create the prime factorization for the following numbers: 45, 75, and 30.

c) $45 =$

d) $75 =$

e) $30 =$

f)   Find the GCF. (Hint: A factor must now be shared across all three numbers.)

The idea is basically the same if we add in variables. Below I've made the prime factorization for $12x^2$ and $15x^3$. Circle the common factors. (Hint: x's are now factor's too.)

$$12x^2 = 2 \cdot 2 \cdot 3 \cdot x \cdot x$$

$$15x^3 = 5 \cdot 3 \cdot x \cdot x \cdot x$$

g) What is the GCF? (Be sure to multiply the x's back together.)

Next, find the GCF of $12x^2y^3$ and $30x^3y^5$

$$12x^2y^3 = 2 \cdot 2 \cdot 3 \cdot x \cdot x \cdot y \cdot y \cdot y$$

$$30x^3y^5 = 2 \cdot 3 \cdot 5 \cdot x \cdot x \cdot x \cdot y \cdot y \cdot y \cdot y \cdot y$$

h) What is the GCF?

Try some that are a bit tricky: $3(x+1)$ and $5(x+1)$. Start by making the prime factorization.

(Hint: Be careful $x+1$ is all one number because it is joined together by addition.)

i) $3(x+1) =$

j) $5(x+1) =$

k) What is the GCF?

$8(x+3)$ and $20(x+3)$. Start by making the prime factorization.

l) $8(x+3) =$

m) $20(x+3) =$

n) What is the GCF?

o)   As always, we can extend the idea. Put a line under each of the factors in the following:

$$5(x + 2)$$

If we multiplied, we would get: $5x + 10$. But instead of distributing, suppose we wanted to go the other direction. We could pull out a GCF and we would get back to where we started.

$$5x + 10 = 5(x + 2)$$

Here's the point. If all the terms of a polynomial share a GCF, we can pull out (divide out) that GCF. In essence, it is like reverse distributing. By doing so, we have turned the polynomial into factors. Factor the following by pulling out a GCF. I've done the first one for you.

$$10x^2 - 15x = 5x(x - 3)$$

p) $3y^3 + 6y^2$

q) $14x^2y^2 - 28x^2y + 21xy$

Sometimes your GCF takes everything from a term. You must leave something behind, so you put a 1. Try these:

r) $12a^3 + 8a^2 + 4a$

s) $x^2y^2 - 5x^3y^3 + 2x^4y^2$

137

t)   Pull out a GCF of -1 from the following. All of the terms left behind will change signs.

$$-x + 5$$

And if your first term begins with a negative, it makes any later math easier if we start by taking out the negative as part of the GCF.

u) $-3y^3 + 6y^2$

v) $-2a^3 + 18a^2 - 4a$

Lesson Eighteen: Additional Exercises

Find the Greatest Common Factor of the following numbers:

a)   15, 45

b)   42, 98

c)   126, 210

d) $21x^2y^3$, $49xy^2$

e) $12(x-1)$, $16(x-1)$

Factor out the GCF for the following:

f) $4x^2 + 24x$

g) $32x^2y^2 - 2xy$

h) $10r^2s + 20rs + 15s$

i) $6a^2 + 3a$

j) $9x^4y^3z^2 - 18x^3y^2z^2 - 9x^2y^2z^2$

k) $-5r^3 + 10r^2$

l) $-14x^5 - 28x^4 + 7x^3$

Lesson Eighteen: Answers

a) 3

b) 6

c) $45: 3 \cdot 3 \cdot 5$

d) $75: 5 \cdot 5 \cdot 3$

e) $30: 2 \cdot 3 \cdot 5$

f) 15

g) $3x^2$

h) $6x^2 y^3$

i) $3 \cdot (x + 1)$

j) $5 \cdot (x + 1)$

k) $(x + 1)$

l) $2 \cdot 2 \cdot 2 \cdot (x + 3)$

m) $2 \cdot 2 \cdot 5 \cdot (x + 3)$

n) $4(x + 3)$

o) factors: $5, x + 2$

p) $3y^2(y + 2)$

q) $7xy(2xy - 4x + 3)$

r) $4a(3a^2 + 2a + 1)$

s) $x^2 y^2(1 - 5xy + 2x^2)$

t) $-1(x - 5)$

u) $-3y^2(y - 2)$

v) $-2a(a^2 - 9a + 2)$

Lesson Eighteen: Additional Exercises Answers

a) 15

b) 14

c) 42

d) $7xy^2$

e) $4(x-1)$

f) $4x(x+6)$

g) $2xy(16xy-1)$

h) $5s(2r^2+4r+3)$

i) $3a(2a+1)$

j) $9x^2y^2z^2(x^2y-2x-1)$

k) $-5r^2(r-2)$

l) $-7x^3(2x^2+4x-1)$

Underline the two factors in the following:

a) $2(x - 2)$

b) $7(y + 3)$

Now, pull out the GCF from these:

c) $3x(x - 2) + 5(x - 2)$

d) $11y(y + 3) + 2(y + 3)$

e) $5a(a - 7) - 2(a - 7)$

f) $4m(m + 2) - (m + 2)$

In this activity, we are going to learn a factoring method called grouping. I will walk you through the idea here. Pull out a GCF from the following:

g) $x^2 + 2x$

h) $3x + 6$

i) Now I've combined everything into one polynomial. Pull out a GCF from the first two terms and pull out a GCF from the last two terms.

$$\underline{x^2 + 2x} + \underline{3x + 6}$$

j) Although it looks strange, your new polynomial has a GCF of $(x + 2)$. In the space below, show how you could pull it out to get the following: $(x + 2)(x + 3)$.

Let's do it again. Pull out a GCF from the following:

k) $y^2 + 4y$

l) $3y + 12$

m) Again, I've combined everything into one polynomial. Pull out a GCF from the first two terms and pull out a GCF from the last two terms.

$$\underline{y^2 + 4y} + \underline{\underline{3y + 12}}$$

n) This rearranged polynomial now has a GCF of $(y + 4)$. In the space below, show how you could pull it out to get the following: $(y + 4)(y + 3)$.

Try some on your own. Always pull out the GCF of the first two terms and then of the last two terms. If you have done it correctly, you will have a matching binomial which then becomes a GCF.

o) $x^2 + 5x + 3x + 15$

p) $x^2 - 6x + 4x - 24$

In this one, the rear of the two terms starts with a negative, so pull out a negative GCF.

q) $y^2 + 2y - 6y - 12$

Here, remember, if you pull out a GCF, you can't leave a spot empty. You put a 1 in its place.

r) $x^2 + x + 6x + 6$

<u>Lesson Nineteen: Additional Exercises</u>

Factor out the GCF:

a) $2x(x+1) + 5(x+1)$

b) $11y(2y-3) - 2(2y-3)$

Factor by grouping:

c) $3x^2 + 6x + 4x + 8$

d) $5x^2 - 15x + 3x - 9$

e) $7y^2 + 14y - 5y - 10$

f) $6r^2 + 9r + 8r + 12$

g) $2x^2 + 2x + x + 1$

h) $15y^2 - 12y - 20y + 16$

a) factors: $2, (x - 2)$

b) factors: $7, (y + 3)$

c) $(x - 2)(3x + 5)$

d) $(y + 3)(11y + 2)$

e) $(a - 7)(5a - 2)$

f) $(m + 2)(4m - 1)$

g) $x(x + 2)$

h) $3(x + 2)$

i) $x(x + 2) + 3(x + 2)$

j) $(x + 2)(x + 3)$

k) $y(y + 4)$

l) $3(y + 4)$

m) $y(y + 4) + 3(y + 4)$

n) $(y + 3)(y + 4)$

o) $(x + 3)(x + 5)$

p) $(x + 4)(x - 6)$

q) $(y - 6)(y + 2)$

r) $(x + 6)(x + 1)$

Lesson Nineteen: Additional Exercises Answers

a) $(x + 1)(2x + 5)$

b) $(2y - 3)(11y - 2)$

c) $(3x + 4)(x + 2)$

d) $(5x + 3)(x - 3)$

e) $(7y - 5)(y + 2)$

f) $(3r + 4)(2r + 3)$

g) $(2x + 1)(x + 1)$

h) $(3y - 4)(5y - 4)$

Below, I have multiplied out a binomial.

$$(x + 5)(x + 6) = x^2 + 5x + 6x + 30$$

a) Combine the like terms and write out the final trinomial.

Here I have multiplied another.

$$(y - 7)(y - 2) = y^2 - 7y - 2y + 14$$

b) Combine the like terms and write out the final trinomial.

c) Multiply out this one on your own. Be sure to combine the like terms.

$$(x - 3)(x + 5) =$$

Factoring polynomials is nothing more than recognizing how we multiplied and then going backward.

$$x^2 \boxed{+ 5x + 6x} + 30$$
$$\wedge$$
$$5*6$$

    1. What two numbers multiplied to get the 30?
    2. And add to get $11x$?

$$y^2 \boxed{- 7y - 2y} + 14$$
$$\wedge$$
$$-7*-2$$

    1. What two numbers multiplied to get the 14?
    2. And add to get $-9y$?

Factor the following: (What two numbers did we multiply to get the last number and add to get the middle?)

d) $x^2 + 10x + 16 = (x+\underline{\quad})(x+\underline{\quad})$

e) $x^2 + 10x + 25 = (x+\underline{\quad})(x+\underline{\quad})$

f) $y^2 - 7y + 12 = (y-\underline{\quad})(y-\underline{\quad})$

On this next one, remember that the signs of the numbers matter. What two numbers multiply to get -18 and add to get -3? One number will need to be positive and one will need to be negative.

g) $$y^2 - 3y - 18 = (y-\underline{\quad})(y+\underline{\quad})$$

Try these on your own.

h) $x^2 + 12x + 20 =$

i) $x^2 - 12x + 35 =$

j) $x^2 + 6x - 7 =$

<u>Lesson Twenty: Additional Exercises</u>

Factor the following:

a) $x^2 + 3x - 10$

b) $y^2 - 12y + 32$

c) $r^2 + 15r + 56$

d) $x^2 + 2x + 1$

e) $y^2 - 7y + 6$

f) $m^2 - m - 6$

g) $x^2 - 16x + 63$

h) $v^2 + 6v - 27$

i) $y^2 - 10y + 16$

j) $x^2 + 12x + 36$

Lesson Twenty: Answers

a) $x^2 + 11x + 30$

b) $y^2 - 9y + 14$

c) $x^2 + 2x - 15$

d) $(x + 8)(x + 2)$

e) $(x + 5)(x + 5)$

f) $(y - 4)(y - 3)$

g) $(y - 6)(y + 3)$

h) $(x + 2)(x + 10)$

i) $(x - 7)(x - 5)$

j) $(x + 7)(x - 1)$

Lesson Twenty: Additional Exercises Answers

a) $(x + 5)(x - 2)$

b) $(y - 8)(y - 4)$

c) $(r + 8)(r + 7)$

d) $(x + 1)(x + 1)$

e) $(y - 6)(y - 1)$

f) $(m - 3)(m + 2)$

g) $(x - 7)(x - 9)$

h) $(v + 9)(v - 3)$

i) $(y - 8)(y - 2)$

j) $(x + 6)(x + 6)$

Lesson 21: Factoring Basic Trinomials with Two Variables

In this activity, we are once again going to factor trinomials. These are going to look quite difficult because they involve two variables. However, I want you to see that it doesn't change much of anything. As you would soon discover, I'm using the same problems from the last activity. The only change is the extra variable. The point is to show you that the process is nearly identical. However, work them without turning back to your previous answers.

Below, I have multiplied out a binomial.

$$(x + 5y)(x + 6y) = x^2 + 5xy + 6xy + 30y^2$$

a)   Combine the like terms and write out the final trinomial.

Here I have multiplied another.

$$(y - 7x)(y - 2x) - y^2 - 7xy - 2xy + 14x^2$$

b)   Combine the like terms and write out the final trinomial.

c)   Multiply out this one on your own. Be sure to combine the like terms.

$$(x - 3y)(x + 5y) =$$

To factor these trinomials, we are still playing the exact same game. What two numbers multiply to get the last term and add to get the middle? Having the extra variable on the end doesn't change anything about the process.

Factor the following:

d) $$x^2 + 10xy + 16y^2 = (x+\underline{\quad}y)(x+\underline{\quad}y)$$

e) $$x^2 + 10xy + 25y^2 = (x+\underline{\quad}y)(x+\underline{\quad}y)$$

f)
$$y^2 - 7xy + 12x^2 = (y - \underline{}\,x)(y - \underline{}\,x)$$

As we saw before, remember that the signs of the numbers matter. What two numbers multiply to get −18 and add to get −3? One number will need to be positive and one will need to be negative.

g)
$$y^2 - 3xy - 18x^2 = (y - \underline{}\,x)(y + \underline{}\,x)$$

Try these on your own.

h)
$$x^2 + 12xy + 20y^2 =$$

i)
$$x^2 - 12xy + 35y^2 =$$

j)
$$x^2 + 6xy - 7y^2 =$$

Lesson Twenty-One: Additional Exercises

Factor the following:

a) $x^2 - 11xy + 30y^2$

b) $r^2 + 12rs + 32s^2$

c) $p^2 + 2pq - 63q^2$

d) $x^2 + 4xy - 60y^2$

e) $m^2 + 24mn + 144n^2$

f) $x^2 + 2xy + y^2$

g) $a^2 - 18ab + 65b^2$

h) $x^2 - 6xy - 72y^2$

Lesson Twenty-One: Answers

a) $x^2 + 11xy + 30y^2$

b) $y^2 - 9xy + 14x^2$

c) $x^2 + 2xy - 15y^2$

d) $(x + 8y)(x + 2y)$

e) $(x + 5y)(x + 5y)$

f) $(y - 3x)(y - 4x)$

g) $(y - 6x)(y + 3x)$

h) $(x + 2y)(x + 10y)$

i) $(x - 7y)(x - 5y)$

j) $(x + 7y)(x - 1y)$

Lesson Twenty-One: Additional Exercises Answers

a) $(x - 6y)(x - 5y)$

b) $(r + 8s)(r + 4s)$

c) $(p + 9q)(p - 7q)$

d) $(x + 10y)(x - 6y)$

e) $(m + 12n)(m + 12n)$

f) $(x + y)(x + y)$

g) $(a - 13b)(a - 5b)$

h) $(x - 12y)(x + 6y)$

Factoring

Lesson 22: The a/c Method for Factoring Trinomials

Below, I'm multiplying out a pair of binomials.

$$(x + 2)(x + 3)$$

$$x^2 + 2x + 3x + 6$$

$$x^2 + 5x + 6$$

Now, I've written it in the opposite direction, going from the trinomial to the factors.

$$x^2 + 5x + 6$$

$$x^2 + 2x + 3x + 6$$

$$(x + 2)(x + 3)$$

a) Do the same thing. First foil it out, showing all the terms. Then, combine like terms.

$$(x + 4)(x + 2)$$

b) Now, write the same problem in the opposite direction, going from the trinomial to the factors.

Factoring is un-foiling a trinomial. However, going backward should include that middle step, but previously we haven't. Yet, the method we are going to learn in this activity will include it. The method is called the a/c method. Let me show you where the middle step is found, and how we get from the middle step to the final answer.

Let's look again at: $x^2 + 5x + 6$.

The a term is the number in front of x^2 and the c term is the last number. Multiply them together.

$$x^2 + 5x + 6$$

In this problem we get 6. Now, play the same game we did before. What two numbers multiply to get 6 and add to get the 5 in the middle? The answer is 3 and 2. Below, I've split the $5x$ into these two numbers. I haven't done anything wrong. They still add to $5x$. Now, factor by grouping. Take the GCF of

c) the first two terms and the GCF of the last two terms.

$$\underline{x^2 + 2x} + \underline{3x + 6}$$

d) Now, you should have a matching term $(x + 2)$. Pull that out as a GCF.

You now have the factored trinomial. We skip this process when the first term of the trinomial is a 1, because it makes things more time consuming. But, if we are really "un-FOIL-ing" a trinomial, we should include the entire process. Try one yourself:

- Multiplying $1 \cdot 8 = 8$.
- What two terms multiply to get 8 and add to get 6?
- Make those your middle terms.
- Factor by grouping.

e)
$$x^2 + 6x + 8$$

When the first term of a trinomial is not a 1, this approach which will always work. The a/c method will take the entire trinomial apart one step at a time and ensure a solution. Let's try one:

$$6x^2 + 10x + 4$$

- We multiply $a \cdot c$ and get 24.
- What two numbers multiply to get 24 and add to get 10? 6 and 4.
  We make those the new middle term: $6x^2 + 6x + 4x + 4$
- Factor by grouping to complete the problem.
$$6x^2 + 6x + 4x + 4$$

f)

Again, this method will always work, and so generally saves a great deal of time over a Trial-and-Error approach.

Try some more:

g) $x^2 - 5x + 6$

h) $2x^2 + 1x - 10$

In this next problem, when you go to factor by grouping, it will seem like there is nothing to take from one of the groups. Remember, in those situations, you must take something and so you take a 1.

i) $3x^2 + 19x + 6$

In this next one, when you factor by grouping, the second group will start with a negative. Remember, in those situations, you take the negative out as part of your GCF. I've started it for you. Finish it by

j) factoring by grouping.

$$12x^2 - 6x - 6$$
$$12x^2 + 6x - 12x - 6$$

Try another on your own:

k) $8x^2 + 10x - 3$

Lesson Twenty-Two: Additional Exercises

Factor the following:

a) $2x^2 + 13x + 15$

b) $3y^2 + 16y - 35$

c) $20p^2 - 11p - 4$

d) $25m^2 + 30m + 9$

e) $28x^2 + 23x - 15$

f) $4y^2 - 19y - 5$

g) $63a^2 - 2a - 48$

h) $6x^2 + xy - 15y^2$

<u>Lesson Twenty-Two: Answers</u>

a) $x^2 + 6x + 8$

b) $x^2 + 4x + 2x + 8 = (x + 4)(x + 2)$

c) $x(x + 2) + 3(x + 2)$

d) $(x + 2)(x + 3)$

e) $(x + 4)(x + 2)$

f) $(6x + 4)(x + 1)$

g) $(x - 2)(x - 3)$

h) $(2x + 5)(x - 2)$

i) $(3x + 1)(x + 6)$

j) $(6x - 6)(2x + 1)$

k) $(4x - 1)(2x + 3)$

<u>Lesson Twenty-Two: Additional Exercises Answers</u>

a) $(2x + 3)(x + 5)$

b) $(3y - 5)(y + 7)$

c) $(5p - 4)(4p + 1)$

d) $(5m + 3)(5m + 3)$

e) $(7x - 3)(4x + 5)$

f) $(4y + 1)(y - 5)$

g) $(9x - 8)(7x + 6)$

h) $(2x - 3y)(3x + 5y)$

Lesson 23: Perfect Square Trinomials

a) I have multiplied out these binomials. Finish by combining like terms.

$$(x - 5)(x - 5) = x^2 - 5x - 5x + 25$$

b)
$$(y + 3)(y + 3) = y^2 + 3y + 3y + 9$$

Multiply the following binomials:

c) $(y - 9)(y - 9)$
d) $(m + 7)(m + 7)$

e) When you are done multiplying this type, you always get a trinomial. What is special about the middle of these trinomials?

f) What causes you to create a middle term like this.?

g) When you are finished multiplying, what is the sign of the third (final) term of the trinomial?

h) There is also something that determines the sign of the middle term. What is it?

As we've seen, factoring is simply following a pattern to un-do a polynomial. You could do the same process that we've done before and you will get the answer. But there is a shortcut which saves so much time it is worth learning. Look at this trinomial.

$$y^2 + 6y + 9$$

Check three things:

1) Does it have a square-root on the first term? Yes. y.
2) Does it have a square-root on the back term? Yes. 3.
3) Is the middle term double those two square roots multiplied together? Yes. $2 \cdot 3 \cdot y = 6y$.

Then, to factor it, write the square roots as two binomials and give them each the sign of the original middle term.

$$(y + 3)(y + 3)$$

i) Try one:

$$x^2 - 8x + 16$$

1) Does it have a square-root on the first term?
2) Does it have a square-root on the back term?
3) Is the middle term double those two square roots multiplied together?

If you answered yes to all of those questions, factor it. (Remember, both terms get the sign of the original middle term.)

Work some more.

j) $x^2 + 10x + 25$

k) $x^2 - 22x + 121$

This next type has an a term, but it doesn't change anything. If it still has a square root, the process remains the same:

l) $9x^2 + 12x + 4$

m) $25x^2 - 40x + 16$

These problems have two variables, but the process remains the same.

n) $x^2 + 10xy + 25y^2$

o) $16x^2 - 24xy + 9y^2$

With so much to factoring, I generally avoid tricky problems. But let me show you one you may see in a textbook. The following polynomial doesn't fit the pattern and so can't be factored.

$$9a^2 + 20x + 16$$

p) Why doesn't it fit?

q) Let's look at one final idea. Write the following number in terms of its prime factors:

$$30 =$$

Notice that there are three factors.

The first step in factoring is always to pull out a GCF. Every time. Sometimes, it may not appear as if you see a factoring pattern, but it may be that you've missed the GCF.

$$4x^2y + 40xy + 100y$$

This has a GCF of $4y$, so pull it out:

$$4y(x^2 + 10x + 25)$$

Then you can finish factoring. But, remember, a GCF is a factor too! So, your answer should include three factors.

$$4y(x + 5)(x + 5)$$

Try some on your own. (First, pull out the GCF.)

r) $x^2y + 6xy + 9y$

s) $2a^2b^3 + 36ab^3 + 162b^3$

Lesson Twenty-Three: Additional Exercises

Factor the following:

a) $x^2 + 12x + 36$

b) $y^2 - 6y + 9$

c) $a^2 + 22a + 121$

d) $r^2 - 16r + 64$

e) $9x^2 + 24xy + 16y^2$

f) $25a^2 - 60ab + 36b^2$

g) $3x^2 + 12x + 12$

h) $9y^3 - 36y^2 + 36y$

i) $16m^3 + 80m^2 + 100m$

j) $98x^3y - 84x^2y + 18xy$

Lesson Twenty-Three: Answers

a) $x^2 - 10x + 25$

b) $y^2 + 6y + 9$

c) $y^2 - 18y + 81$

d) $m^2 + 14m + 49$

e) It is double the two terms of the binomial multiplied together.

f) Foiling the same terms causes two of the middle term.

g) It will always be positive.

h) The sign of the terms you started with.

i) $(x - 4)(x - 4)$

j) $(x + 5)(x + 5)$

k) $(x - 11)(x - 11)$

l) $(3x + 2)(3x + 2)$

m) $(5x - 4)(5x - 4)$

n) $(x + 5y)(x + 5y)$

o) $(4x - 3y)(4x - 3y)$

p) The middle term is not twice of the two terms multiplied.

q) $3 \cdot 2 \cdot 5$

r) $y(x + 3)(x + 3)$

s) $2b^3(a + 9)(a + 9)$

Lesson Twenty-Three: Additional Exercises Answers

a) $(x + 6)(x + 6)$

b) $(y - 3)(y - 3)$

c) $(a + 11)(a + 11)$

d) $(r - 8)(r - 8)$

e) $(3x + 4y)(3x + 4y)$

f) $(5a - 6b)(5a - 6b)$

g) $3(x + 2)(x + 2)$

h) $9y(y - 2)(y - 2)$

i) $4m(2m + 5)(2m + 5)$

j) $2xy(7x - 3)(7x - 3)$

Factoring

Lesson 24: Differences of Squares

Let's learn another factoring shortcut by recalling a pattern. I have multiplied out these binomials.

a) Finish by combining like terms.

$$(x+5)(x-5) = x^2 + 5x - 5x - 25$$

Multiply the following binomials:

b) $(y-9)(y+9)$

c) $(m-7)(m+7)$

d) What keeps happening to the middle?

e) Why does this keep happening?

f) When you are finished multiplying, you are always left with a binomial. What is the sign of this binomial?

Factoring is recognizing the pattern of taking a polynomial in reverse. Here's what is needed for our newest shortcut.

$$x^2 - 9$$

1) Does the first term have a square root? Yes. x.
2) Does the last term have a square root? Yes. 3.
3) Is the sign in between a negative? Yes.

Then, to factor it, write the square roots as two binomials and give them each a different sign.

$$(x - 3)(x + 3)$$

g) Try on your own.

$$y^2 - 16$$

1) Does the first term have a square root?
2) Does the last term have a square root?
3) Is the sign in between a negative?

If you answered yes to all of those questions, factor it. (Remember, both terms get opposite signs.)

Work some more.

h) $x^2 - 25$

i) $x^2 - 121$

This next type has an a term, but it doesn't change anything. If it still has a square root, the process remains the same:

j) $9x^2 - 4$

k) $25x^2 - 16$

These problems have two variables, but the process remains the same.

l) $x^2 - 25y^2$

m) $16x^2 - 9$

These next two reverse the order, but that's okay. Just reverse the way you write it.

$$25 - x^2 = (5 - x)(5 + x)$$

n) Factor: $121 - y^2$

o) Can you give me the prime factorization of 7? Why or why not?

Multiply out each of the following:

p) $(x + 3)(x + 3)$

q) $(x - 3)(x - 3)$

r) $(x + 3)(x - 3)$

s) Notice that we've done each possible sign combination. Do any of these multiply out to get $x^2 + 9$?

If you have a binomial with square roots on the front and square roots on the back, you must have a negative in the middle. Binomials with a positive in the middle can't be factored. There is no combination that can multiply to create it. So, $x^2 + 9$ is a prime number.

t) Give me the prime factorization of 50.

$$50 =$$

This next problem has a higher power. But it still fits the pattern:

$$x^4 - 16$$

1) Does the first term have a square root? Yes. x^2
2) Does the last term have a square root? Yes. 4.
3) Is the sign in between a negative? Yes.

So, we factor and get:$(x^2 - 4)(x^2 + 4)$. But just like when we find a prime factorization, sometimes the
u) numbers can go further. $(x^2 + 4)$ is prime. But $(x^2 - 4)$ can be factored further. In the space below, factor $x^4 - 16$ as far as it will go. Remember, $(x^2 + 4)$ is a factor and must be part of your final answer.

v) Try another. Factor as far as it will go.

$a^4 - b^4$

Finally, as we saw in the last section, pulling out a GCF is always the first step in factoring. These next polynomials fit our pattern (called a perfect square binomial) but you must start by pulling out the GCF. Remember to include the GCF as part of the final factored solution.

w) $x^2y - 100y$

x) $5a^2b^3 - 125b^3$

y) $2x^4 - 32$

<u>Lesson Twenty-Four: Additional Exercises</u>

Factor the following:

a) $x^2 - 25$

b) $y^2 - 64$

c) $9a^2 - 16b^2$

d) $25r^2 - 169s^2$

e) $12x^2 - 300$

f) $147m^2 - 363n^2$

g) $81y^2 + 16$

h) $x^4 - 16$

i) $5y^4 - 80$

j) $a^4 - b^4$

Lesson Twenty-Four: Answers

a) $x^2 - 25$

b) $y^2 - 81$

c) $m^2 - 49$

d) The middle terms drop out.

e) The middle terms have opposite signs.

f) negative

g) $(y - 4)(y + 4)$

h) $(x - 5)(x + 5)$

i) $(x - 11)(x + 11)$

j) $(3x - 2)(3x + 2)$

k) $(5x - 4)(5x + 4)$

l) $(x - 5y)(x + 5y)$

m) $(4x - 3)(4x + 3)$

n) $(11 - y)(11 + y)$

o) No, it is prime.

p) $x^2 + 6x + 9$

q) $x^2 - 6x + 9$

r) $x^2 - 9$

s) No

t) $2 \cdot 5 \cdot 5$

u) $(x^2 + 4)(x - 2)(x + 2)$

v) $(a^2 + b^2)(a - b)(a + b)$

w) $y(x - 10)(x + 10)$

x) $5b^3(a - 5)(a + 5)$

y) $2(x^2 + 4)(x - 2)(x + 2)$

Lesson Twenty-Four: Additional Exercises Answers

a) $(x - 5)(x + 5)$

b) $(y - 8)(y + 8)$

c) $(3a - 4b)(3a + 4b)$

d) $(5r - 13s)(5r + 13r)$

e) $12(x - 5)(x + 5)$

f) $3(7m - 11n)(7m + 11n)$

g) prime

h) $(x - 2)(x + 2)(x^2 + 4)$

i) $5(y - 2)(y + 2)(y^2 + 4)$

j) $(a - b)(a + b)(a^2 + b^2)$

Lesson Twenty-Four: Answers

a) $x^2 - 25$

b) $y^2 - 81$

c) $m^2 - 49$

d) The middle terms drop out.

e) The middle terms have opposite signs.

f) negative

g) $(y - 4)(y + 4)$

h) $(x - 5)(x + 5)$

i) $(x - 11)(x + 11)$

j) $(3x - 2)(3x + 2)$

k) $(5x - 4)(5x + 4)$

l) $(x - 5y)(x + 5y)$

m) $(4x - 3)(4x + 3)$

n) $(11 - y)(11 + y)$

o) No, it is prime.

p) $x^2 + 6x + 9$

q) $x^2 - 6x + 9$

r) $x^2 - 9$

s) No

t) $2 \cdot 5 \cdot 5$

u) $(x^2 + 4)(x - 2)(x + 2)$

v) $(a^2 + b^2)(a - b)(a + b)$

w) $y(x - 10)(x + 10)$

x) $5b^3(a - 5)(a + 5)$

y) $2(x^2 + 4)(x - 2)(x + 2)$

a) $(x - 5)(x + 5)$

b) $(y - 8)(y + 8)$

c) $(3a - 4b)(3a + 4b)$

d) $(5r - 13s)(5r + 13r)$

e) $12(x - 5)(x + 5)$

f) $3(7m - 11n)(7m + 11n)$

g) prime

h) $(x - 2)(x + 2)(x^2 + 4)$

i) $5(y - 2)(y + 2)(y^2 + 4)$

j) $(a - b)(a + b)(a^2 + b^2)$

Factoring

Lesson 25: Sum or Difference of Cubes

In this activity, we will see that there is also a pattern for binomials which have cubes on the front and back.

a) Finish multiplying the problem which I have started below. I'd recommend foiling the last two binomials. Then you will need to multiply your answer by the remaining $(x - 3)$.

$$(x - 3)(x - 3)(x - 3)$$

If you did this correctly, you should have gotten a very complicated answer. Here's the point, look at this binomial:

$$x^3 - 27$$

It has cube roots on the first term and on the last term, but it doesn't turn into this:

$$x^3 - 27 \neq (x - 3)(x - 3)(x - 3)$$

The reality is a bit more unfortunate. To factor a binomial which has cube roots on the first term and the last term, you need to know two things:

1) The pattern $(a\ b)(a^2\ ab\ b^2)$.
2) And the acronym: S.O.A.P.

Let's walk through a problem.

$$x^3 - 27$$

1) Does it have a cube root on the front? Yes. x.
2) Does it have a cube root on the back? Yes. 3.

Note: These can have a positive or a negative in between!

Now, we will fit this to our pattern $(a\ b)(a^2\ ab\ b^2)$. a equals the cube root of the first term. b equals the cube root of the last term.

So:

$$a = x$$

$$b = 3$$

Then we put those into our pattern:

$a = x$

$b = 3$

$a^2 = x^2$

$ab = 3x$

$b^2 = 9$

$$(x\ 3)(x^2\ 3x\ 9)$$

But we also need signs between the terms. That's where S.O.A.P. comes in. The acronym stands for:

S = Same

O = Opposite

A = Always

P = Positive

To begin, we put in the same sign from the original problem. $x^3 - 27$ The sign was negative, so our first sign is negative.

$$(x - 3)(x^2\ 3x\ 9)$$

Opposite means the next sign is always the opposite of the first.

$$(x - 3)(x^2 + 3x\ 9)$$

And Always Positive means that the last sign is always a positive.

$$(x - 3)(x^2 + 3x + 9)$$

b) Try one.

$$x^3 - 8$$

180

c) Try another. Notice that the sign in between is a positive this time.

$$y^3 + 64$$

d) Nothing changes if both the terms are variables:

$$x^3 - y^3$$

$a = x$

$b = y$

$a^2 = x^2$

$ab = xy$

$b^2 = y^2$

Finish this problem.

e) This next problem is more complicated, but the process is still the same.

$$27r^3 - 8s^3$$

$a = 3r$

$b = 2s$

$a^2 = 9r^2$ (Notice that all of a has been squared.)

$ab = 6rs$ (Notice that all of a and b have been multiplied.)

$b^2 = 4s^2$ (Notice that all of b has been squared.)

Finish this problem.

f) Finally, if there is a GCF, always pull it out first.

$$2x^3z + 54y^3z$$

Factor. Be sure to include the GCF as part of the answer.

Lesson Twenty-Five: Additional Exercises

Factor the following:

a) $x^3 - 125$

b) $y^3 + 64$

c) $a^3 - 216$

d) $8r^3 - 1$

e) $27x^3 + 8$

f) $m^3 + n^3$

g) $64x^3 - y^3$

h) $216r^3 + 343s^3$

Lesson Twenty-Five Answers

a) $x^3 - 9x^2 + 27x - 27$

b) $(x - 2)(x^2 + 2x + 4)$

c) $(y + 4)(y^2 - 4y + 16)$

d) $(x - y)(x^2 + xy + y^2)$

e) $(3r - 2s)(9r^2 + 6rs + 4s^2)$

f) $2z(x + 3y)(x^2 - 3xy + 9y^2)$

Lesson Twenty-Five: Additional Exercises Answers

a) $(x - 5)(x^2 + 5x + 25)$

b) $(y + 4)(y^2 - 4t + 16)$

c) $(a - 6)(a^2 + 6a + 36)$

d) $(2r - 1)(4r^2 + 2r + 1)$

e) $(3x + 2)(9x^2 - 6x + 4)$

f) $(m + n)(m^2 - mn + n^2)$

g) $(4x - y)(16x^2 + 4xy + y^2)$

h) $(6r + 7s)(36r^2 - 42rs + 49s^2)$

Lesson 26: A Strategy for Factoring

Factoring gets more difficult when all the different patterns are mixed together. When we are presented with that situation, there is a course of action we should follow:

1) Check first for a Greatest Common Factor.
2) Check for square roots on the front and back. If you have them, you may be dealing with:
   a. A perfect square trinomial.
   b. A perfect square binomial.
3) Check for cube roots on the front and back. You may have a difference of cubes.
4) If there are four terms, try factoring by grouping.
5) If none of those apply, you have a traditional factoring problem:
   a. If the leading coefficient is 1, you can find two numbers to multiply to get the last term and add to get the middle term.
   b. If the leading coefficient isn't 1, use the a/c method.

Use the approach above to factor the following:

a) $x^2 - 2x - 35$

b) $3x^2 + 16x - 12$

c) $x^2 - 8x + 16$

d) $x^2 - 9$

e) $x^2y - 15xy + 56y$

f) $x^3 - 27y^3$

g) $16x^4 + 16x^3 + 4x^2$

h) $x^2 - 6x - 27$

i) $2x^2 + 11x + 15$

j) $x^4 - 81$

k) $x^5 + x^3y^2$

l) $x^2 + x + 3x + 3$

Lesson Twenty-Six: Additional Exercises

Factor the following:

a) $6x^2 - x - 2$

b) $27y^3 - 8$

c) $4y^2 - 25$

d) $x^2 + x - 72$

e) $7y^2 + 70y + 147$

f) $15x^3y^2 - 25x^2y + 50xy^2$

g) $m^2 - 26m + 169$

h) $24x^4 + 6x^3 - 63x^2$

i) $1 - r^3$

j) $a^4 - 16b^4$

Lesson Twenty-Six: Answers

a) $(x + 5)(x - 7)$

b) $(3x - 2)(x + 6)$

c) $(x - 4)(x - 4)$

d) $(x - 3)(x + 3)$

e) $y(x - 7)(x - 8)$

f) $(x - 3y)(x^2 + 3xy + 9y^2)$

g) $4x^2(2x + 1)(2x + 1)$

h) $(x - 9)(x + 3)$

i) $(2x + 5)(x + 3)$

j) $(x - 3)(x + 3)(x^2 + 9)$

k) $x^3(x^2 + y^2)$

l) $(x + 3)(x + 1)$

Lesson Twenty-Six: Additional Exercises Answers

a) $(3x - 2)(2x + 1)$

b) $(3y - 2)(9y^2 + 6y + 4)$

c) $(2y - 5)(2y + 5)$

d) $(x + 9)(x - 8)$

e) $7(y + 7)(y + 3)$

f) $5xy(3x^2y - 5x + 10y)$

g) $(m - 13)(m - 13)$

h) $3x^2(4x + 7)(2x - 3)$

i) $(1 - r)(1 + r + r^2)$

j) $(a - 2b)(a + 2b)(a^2 + 4b^2)$

Relations and Functions

So far, we have seen that Algebra is solving for and working with missing numbers. In this final section, things will appear to be similar, but in reality, they are rather different. Equal signs will return and a second variable will be introduced. For instance:

$$x + y = 9$$

Previously, when we saw an equal sign, we would solve to find the value of the variable. But here, the second variable makes that impossible. Instead, we encounter a new idea called relations.

In the equation $x + y = 9$, the variables x and y have a relationship. Just as in any relationship, this means that they interact with each other. Together, they must always add to 9. But there are multiple ways in which this can be done. The table below shows some different possibilities:

x	y
1	8
2	7
3	6
4	5
5	4
6	3
7	2
8	1

Each x has a partner value with y in which they add to 9. And, in fact, if we consider decimals, there are unlimited combinations of these partnered values. This is a new idea for us. Previously, a variable would represent a single missing number. Now, the variable can be a placeholder for any number in the world, provided that it is in the right relationship with its partner.

Here, in our final section, we will examine these relationships, which mathematicians simply call relations. Interestingly, these relations can be drawn and this can result in shapes. There are many shapes formed by relations but the simplest and easiest to understand is a line. So, we will begin there and we will end by discussing a special kind of relation known as a function.

Lesson 27: The Coordinate Plane

In algebra, one of our major ideas is graphing. The basis of the concept is like a treasure hunt.

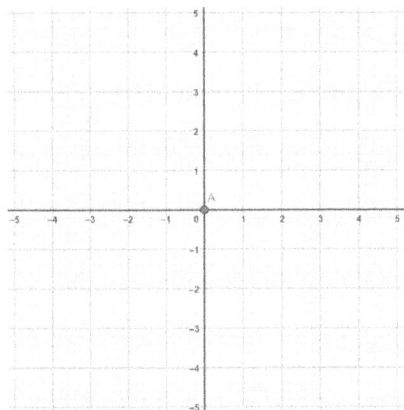

Imagine a treasure has been buried in a field. To find it, the field has been divided into a grid.

You start at the middle (called the origin). The instructions tell you where to go.

The first instruction tells you whether to go right or left. (Positive is right. Negative is left.) The second tells you whether to go up or down. (Positive is up. Negative is down.)

Here is an example: $(-2, 3)$

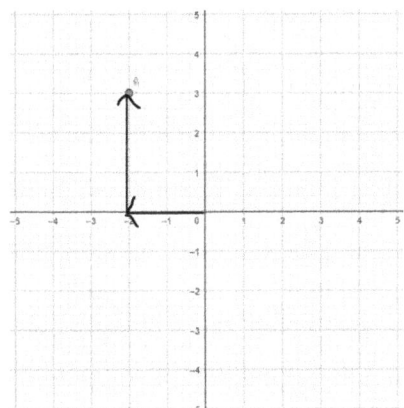

The first instruction was -2. So, I went left 2.

The second instruction was 3. So, I went up 3.

Try some on your own. Start at the origin (0,0) for each.

a) $(3, -2)$ b) $(-1, 4)$ c) $(-2, -4)$

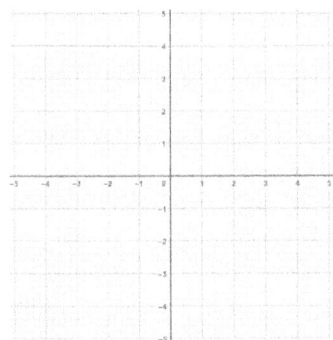 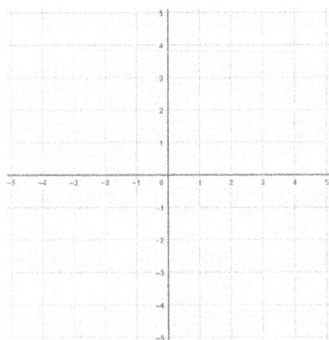 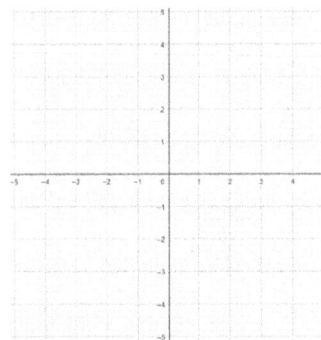

This system divides the field into four boxes called Quadrants. The boxes go counterclockwise.

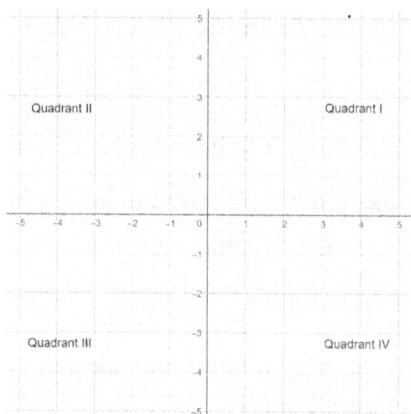

The point $(2, -3)$ would be in quadrant IV.

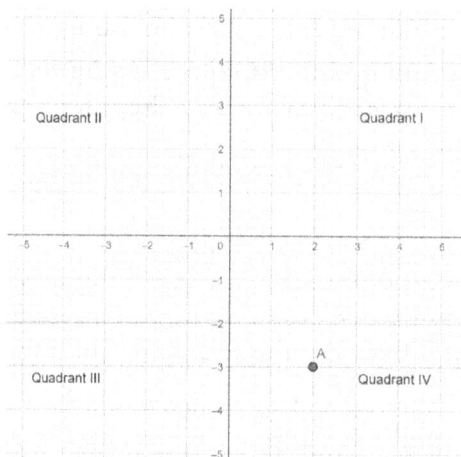

Graph the following points *and* state which quadrant they would be in:

d) $(-2, 3)$ e) $(1, 3)$ f) $(-1, -3)$

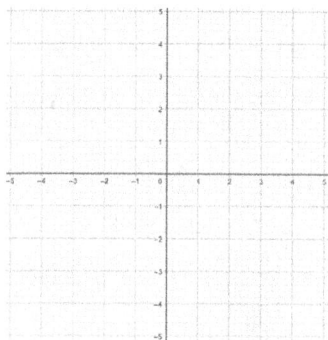   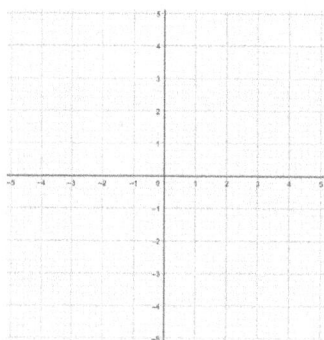   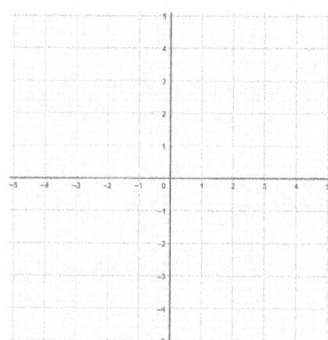

A point isn't always in a quadrant. They can also be on one of the two major lines.

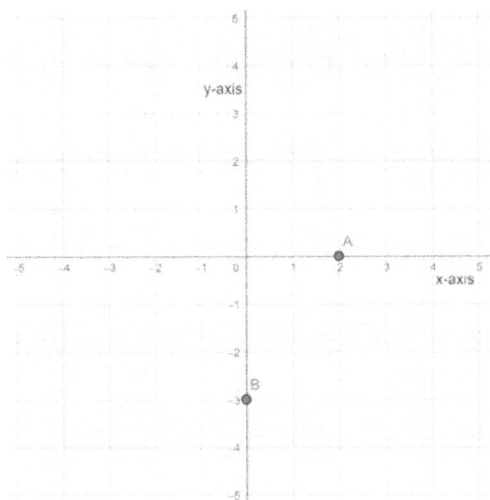

The right and left line is called the x-axis.

The up and down line is called the y-axis.

Point A is on the x-axis. Point B is on the y-axis.

Which axis would the following points be on:

g) (3, 0)   h)   (0, 4)   i)   (−1, 0)

In this system, a point is called an ordered pair. The first instruction, right or left, is called the x-coordinate. The second instruction, up or down, is called the y-coordinate. So: (x, y).

For the following points, state the x and y coordinates:

$(−2, 3)$

j)   x-coordinate:

y-coordinate:

$(−1, −3)$

k)   x-coordinate:

y-coordinate:

$(8, 12)$

l)   x-coordinate:

    y-coordinate:

$(0, 0)$

m)   x-coordinate:

    y-coordinate:

n)   The final point has a special name. What was its name?

Lesson Twenty-Seven: Additional Exercises

Plot the following points and give the quadrant. If it isn't in a quadrant, give the axis:

a) $(-3, 4)$

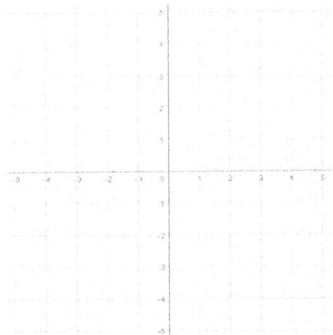

b) $(5, -2)$

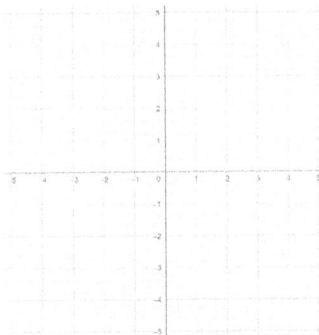

c) $(0, -4)$

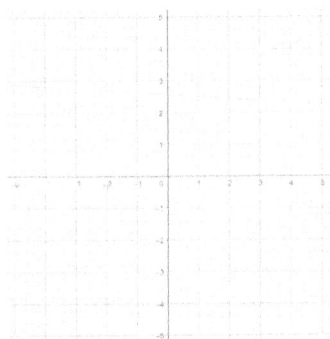

d) $(2, 1)$

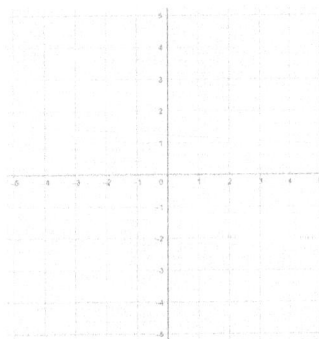

e) $(3, 0)$

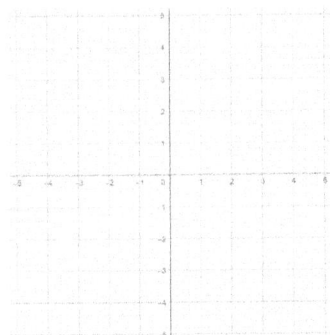

f) $(0, 0)$

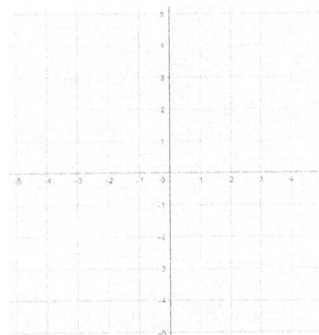

195

For the following points, state the x and y coordinates:

g) $(7, -3)$

x-coordinate:

y-coordinate:

h) $(-5, -13)$

x-coordinate:

y-coordinate:

i) $(11, 12)$

x-coordinate:

y-coordinate:

j) $(3, -15)$

x-coordinate:

y-coordinate:

Lesson Twenty-Seven: Answers

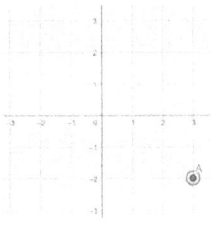

a)

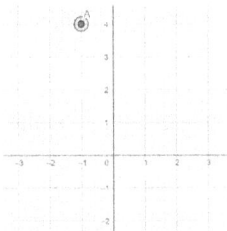

b)

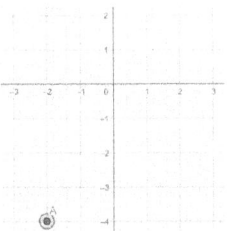

c)

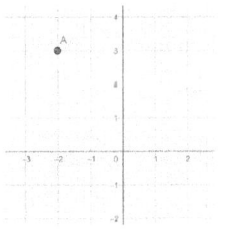

d)   Quadrant II

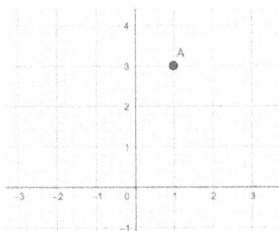

e)   Quadrant I

f)   Quadrant III

g) x-axis

h) y-axis

i) x-axis

j) x-coordinate: -2

y-coordinate: 3

k) x-coordinate: -1

y-coordinate: -3

l) x-coordinate: 8

y-coordinate: 12

m) x-coordinate: 0

y-coordinate: 0

n) origin

Lesson Twenty-Seven: Additional Exercises Answers

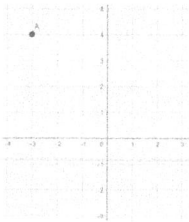

a)   Quadrant II

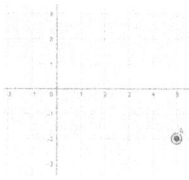

b)   Quadrant IV

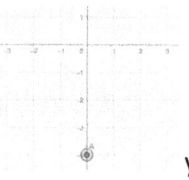

c)   y-axis

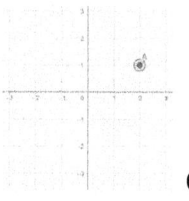

d)   Quadrant I

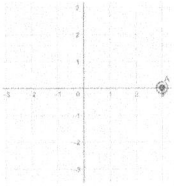

e)   x-axis

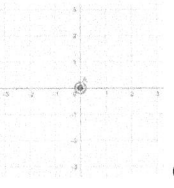

f)   origin

g)

x-coordinate: 7

y-coordinate: -3

h)

x-coordinate: -5

y-coordinate: -13

i)

x-coordinate: 11

y-coordinate: 12

j)

x-coordinate: 3

y-coordinate: -15

Graphing is one of the main concepts of algebra. And it is built on the ideas of ordered pairs and relations. We saw ordered pairs in the last activity. Today we want to look at relations. Here's an example.

$$x + y = 8$$

The equal sign makes this an equation. But because there are two variables, this also means something else. The two variables are in relationship with one another. If you add x and y you get 8. Yet since these are variables, there are lots of ways to do it. I showed you the first one. Finish the table.

a)

x	y
3	5
5	
2	
-1	
-4	

There are more than this. In fact, there are an infinite number of combinations. Each working combination is called a solution to that equation, and they create an ordered pair. I've plotted the first ordered pair $(3, 5)$ on the graph below. Add the rest of the ordered pairs from the table.

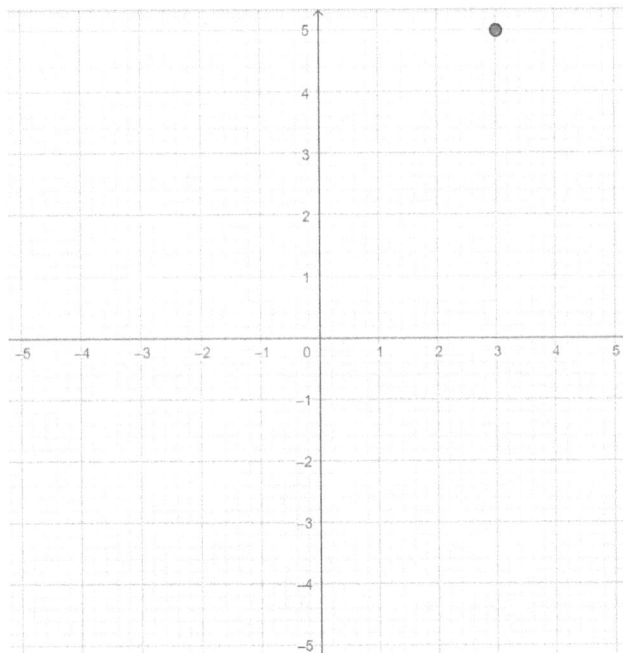

b)

This relationship (just called a relation in math) between x and y makes a line. Connect the points from the table which you graphed. (If your points are correct than you should have a line.) But there are infinite possible points and so we extend the line in both directions and add arrows on the ends. (Extend your line and add the arrows.)

Here is another relation. This one is a little bit more complicated. You may need your algebra skills.

$$2x + y = 8$$

I'm going to start with $x = 3$ and find the y which is his partner.

$$2(3) + y = 8$$

$$6 + y = 8$$

$$y = 2$$

c) I've done the first two ordered pairs. Finish the table.

X	y
3	2
2	4
5	
1	
-1	

d) Graph the ordered pairs and make a line.

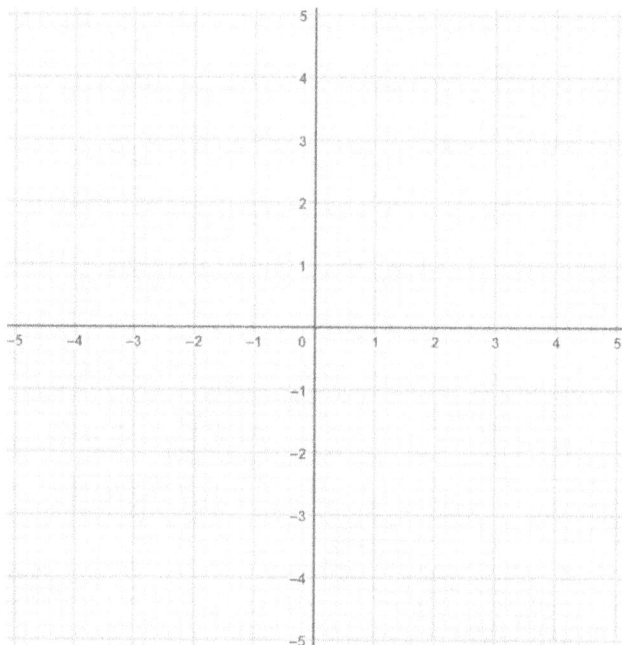

e) Here's another. The relation is presented differently; however, it is still the same idea.

$$y = 4x - 3$$

x	y
0	
-1	
2	

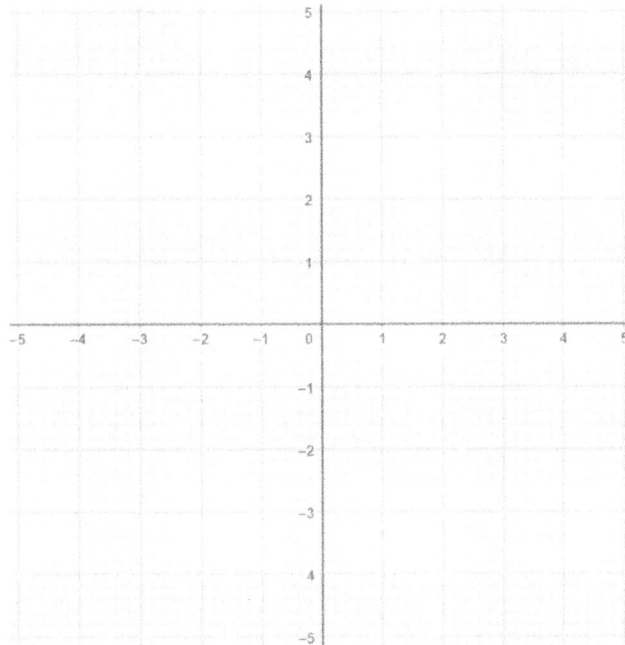

Because the line goes on forever, every value of x has a partner value of y. So, you could actually pick any value of x you wanted and then find the partner y. The problem is that some of them turn out to be fractions which are difficult to graph. If you are picking your own points, make choices which cause the math to be easy. (For instance, I always use 0.) Three points are all that is required to graph a line.

$$y = 3x - 4$$

f) Pick your own points and graph the line:

x	y

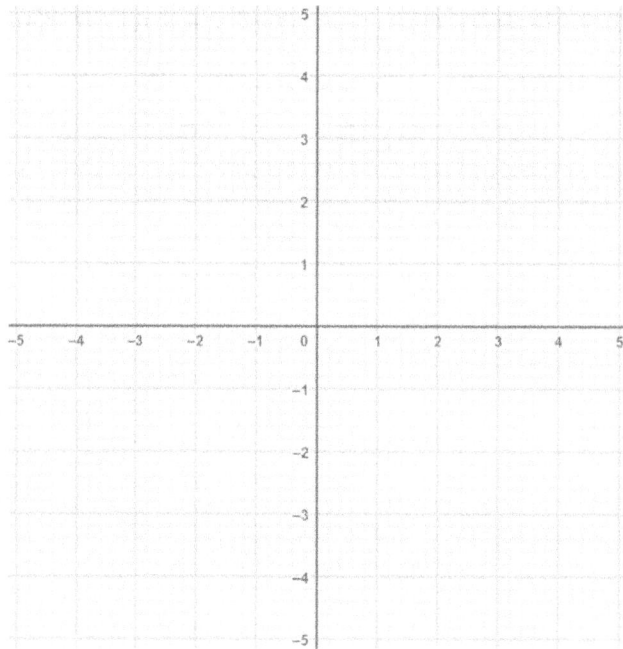

Finally, any ordered pair which makes the relation true is called a solution. If you graphed the line, solutions are ordered pairs which are on the line.

Look at this relation:

$$3x + 2y = 12$$

g) Circle any of the following ordered pairs which are solutions:

$(0, 6)$

$(-2, 8)$

$(-4, 12)$

$(6, -3)$

$(10, 9)$

a) Complete the table below and use your points to graph the line $y + 2x = 4$

x	y
1	
2	
3	
-1	
-2	

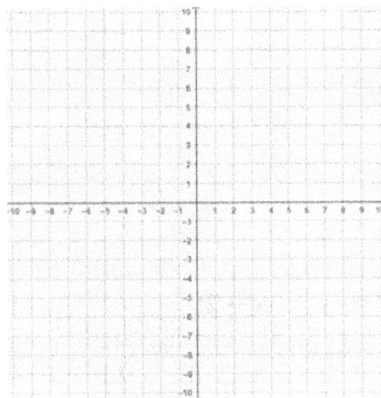

b) Complete the table below and use your points to graph the line $y - 3x = -2$

x	y
1	
2	
3	
-1	
-2	

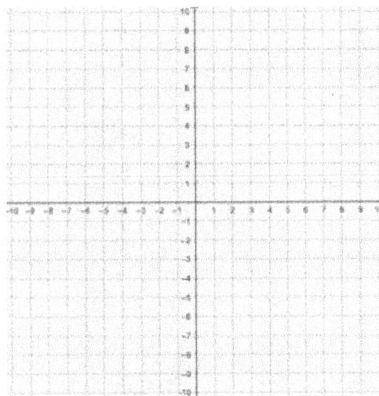

c) Pick your own points and graph the line $2x + y = 4$:

x	y

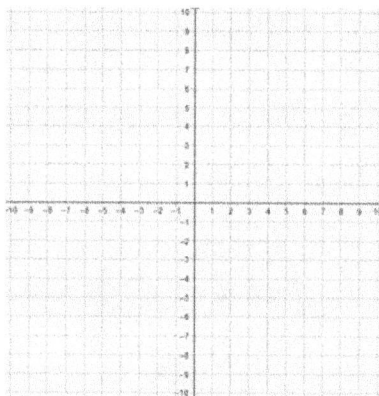

d) Pick your own points and graph the line $y = 3x - 2$:

x	y

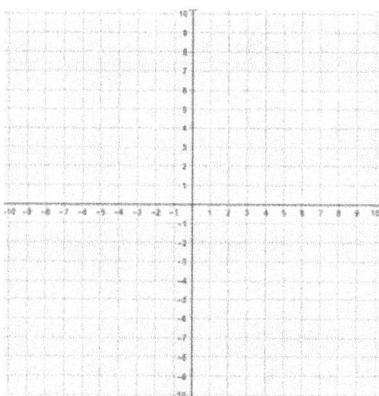

Look at this relation:

$$5x - 3y = 14$$

e) Circle any of the following ordered pairs which are solutions:

$(-2, -3)$

$(-2, -8)$

$(0, -2)$

$(7, 7)$

$(1, -2)$

Look at this relation:

$$y = -x + 5$$

f) Circle any of the following ordered pairs which are solutions:

$(5, 5)$

$(-1, 6)$

$(2, 2)$

$(3, 3)$

$(5, 0)$

Lesson Twenty-Eight: Answers

a)

x	y
3	5
5	3
2	6
-1	9
-4	12

b)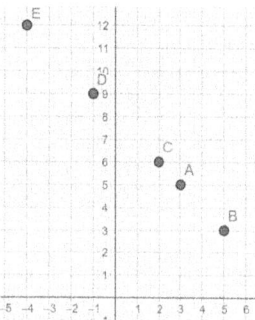

c)

x	y
3	2
2	4
5	-2
1	6
-1	10

d)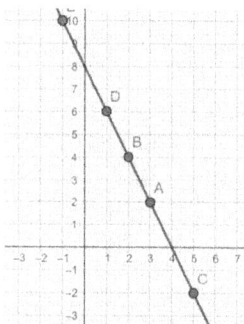

e)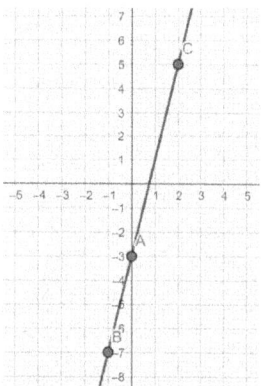

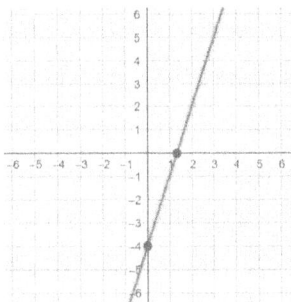

f)

g) a, c, d

Lesson Twenty-Eight: Additional Exercises Answers

a)

x	y
1	2
2	0
3	-2
-1	6
-2	8

b)

x	y
1	1
2	4
3	7
-1	-5
-2	-8

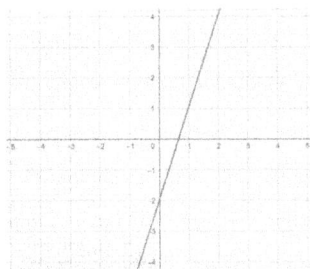

c) Points selected may vary.

x	y
0	4
1	2
-1	6

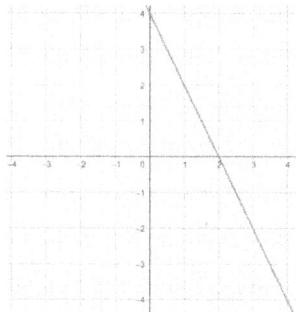

d) Points selected may vary.

x	y
0	-2
1	1
-1	-5

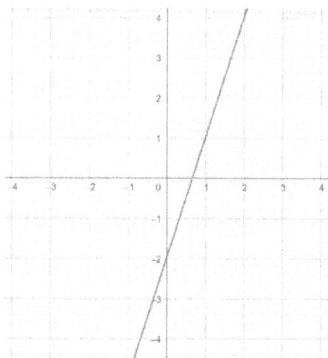

e) $(-2, -8), (7, 7)$

f) $(-1, 6), (5, 0)$

Below is the graph of a line:

$$y = 3x + 1$$

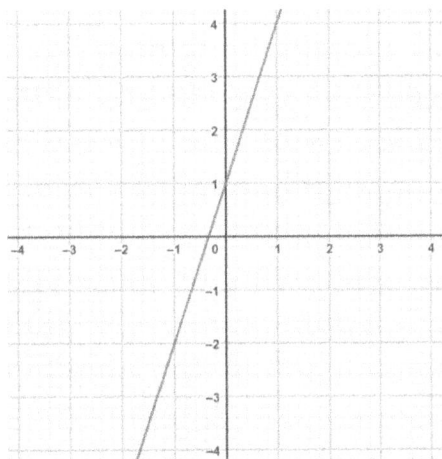

a) An ordered pair is a solution if it is on the line. Circle any of the following which are a solution to the line: $y = 3x + 1$.

   a) $(1, 4)$

   b) $(-1, -1)$

   c) $(-1, -2)$

   d) $(0, 2)$

To graph a line, you need three points. Any values of x will do, but picking easy values helps with the work. With fractions, let x be the same as the denominator.

$$y = \frac{1}{4}x - 3$$

$$y = \frac{1}{4}(4) - 3$$

$$y = 1 - 3$$

$$y = -2$$

b) When your x value is the same as the denominator, it will cancel. Finish the table below and then graph the line.

$$y = \frac{1}{4}x - 3$$

x	y
4	-2
0	
-4	

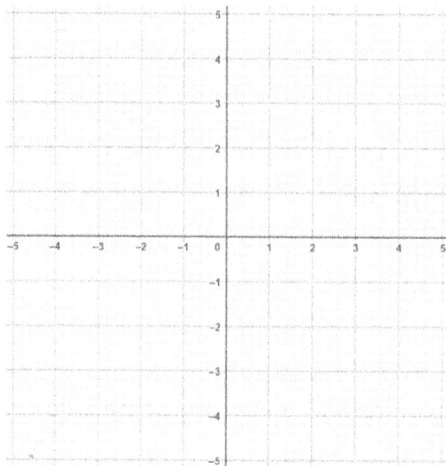

c) Try another: $y = \frac{3}{5}x - 2$

x	y
5	
0	
-5	

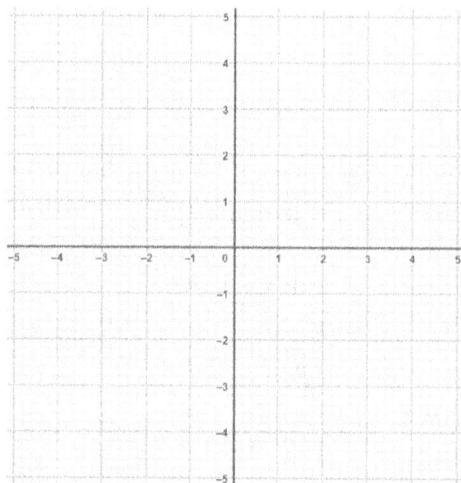

Lesson Twenty-Nine: Additional Exercises

a)   Complete the table and graph the relation $y = \frac{1}{3}x - 2$

x	y
-3	
0	
3	

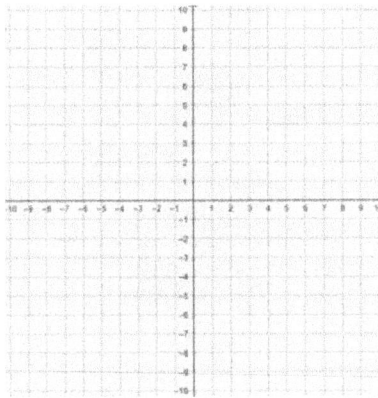

b)   Complete the table and graph the relation $y = -\frac{2}{7}x + 3$

x	Y
-7	
0	
7	

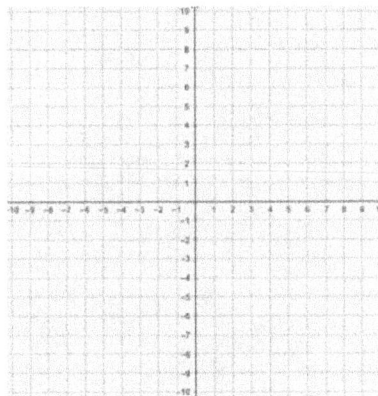

c)　Select values of x, complete the table, and graph the relation $y = \frac{1}{2}x - 1$

x	y

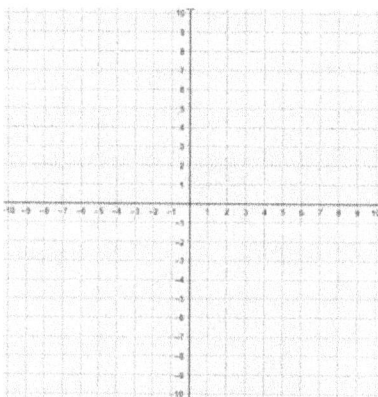

d)　Select values of x, complete the table, and graph the relation $y = -\frac{2}{3}x + 4$

x	y

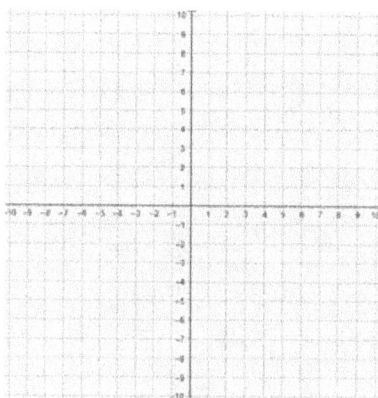

Lesson Twenty-Nine: Answers

a) a, c

b)

x	y
4	-2
0	-3
-4	-4

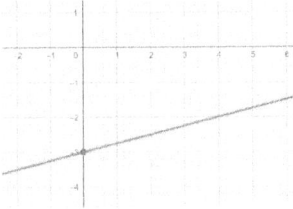

c)

x	y
5	1
0	-2
-5	-5

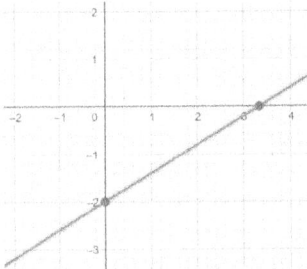

Lesson Twenty-Nine: Additional Exercises Answers

a)

x	y
-3	-3
0	-2
3	-1

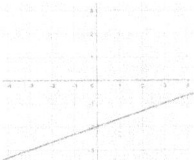

b)

x	y
-7	5
0	3
7	1

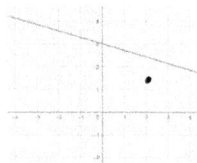

c)

x	y
-2	-2
0	-1
2	1

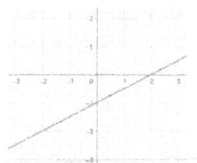

d)

x	y
-3	6
0	4
3	2

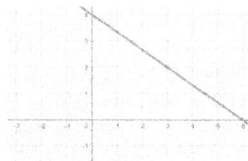

Lesson 30: Graphing Vertical and Horizontal Lines

In this activity, we are going to graph very strange lines. Look at this one:

$$y = 3$$

a) It says that y is always 3. So, no matter what x values we choose, y will be 3. Complete the table and graph the line. (Don't over think it.)

x	y
1	
0	
-1	

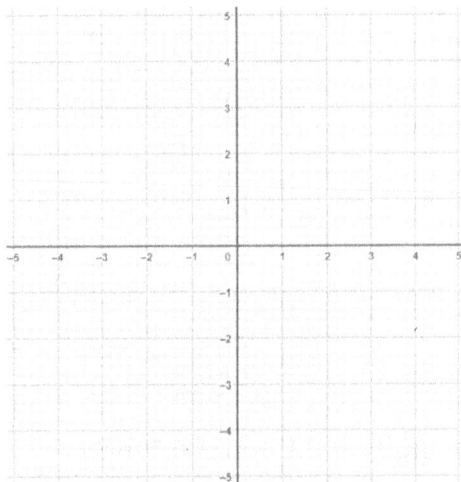

b) Once you see what will happen, you could have done this without the table. Graph $y = -1$. The y's will always just be -1.

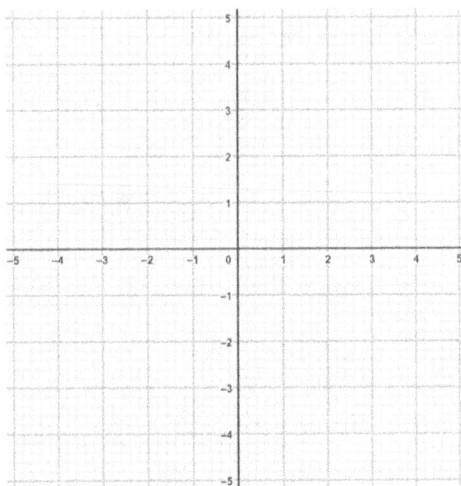

Now, let's try one with just x.

$$x = -2$$

c) Our table is backward, but it works the same. The values of x will always be -2.

x	y
	-1
	0
	1

Graph the line.

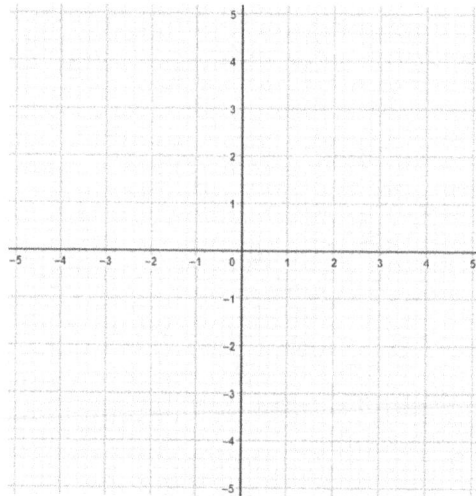

d) Again, the table is unnecessary. Graph the line $x = 1$. (x is always 1.)

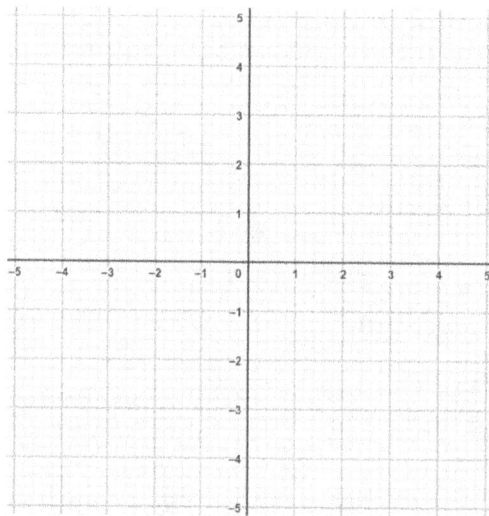

e) If you see the pattern, you know that lines involving $y =$ are always_____.

(Circle the correct answer.)

    a) Horizontal
    b) Vertical

f) Similarly, you know that lines involving $x =$ are always_____. (Circle the correct answer.)

    a) Horizontal
    b) Vertical

Lesson Thirty: Additional Exercises

a) Graph the following lines:

$$y = -2$$

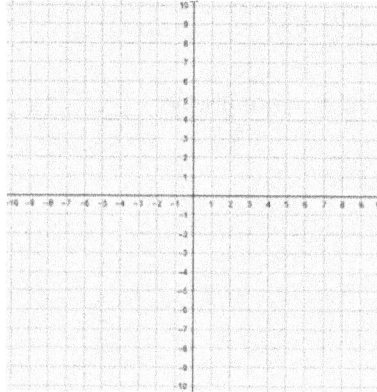

b)

$$x = 5$$

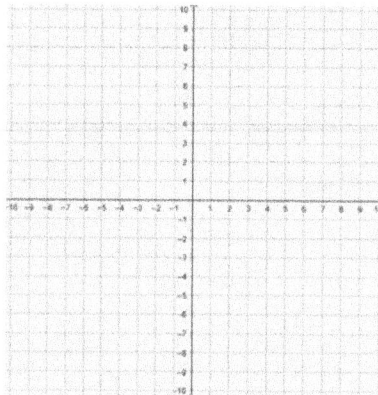

c)

$$x = -1$$

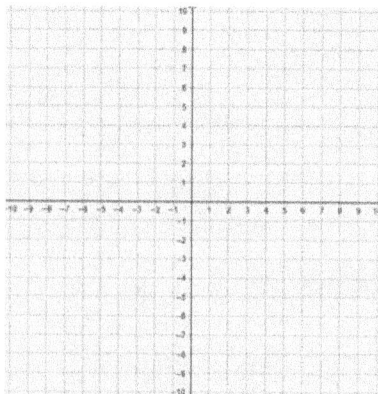

d)

$$y = 4$$

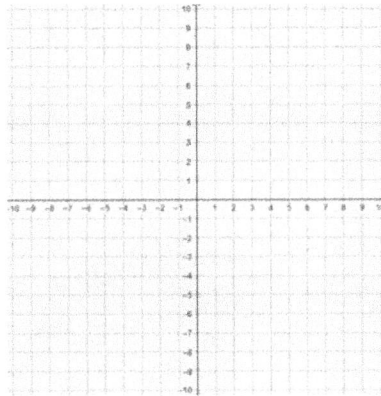

Lesson Thirty: Answers

a)

x	y
1	3
0	3
-1	3

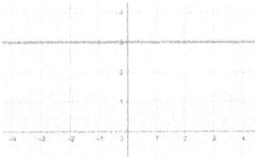

b)

c)

x	y
-2	-1
-2	0
-2	1

d)

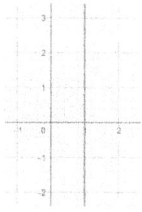

e) horizontal

f) vertical

Lesson Thirty: Additional Exercises Answers

a)

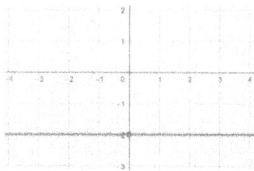

b)

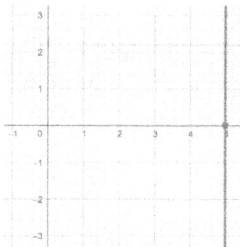

c)

d)

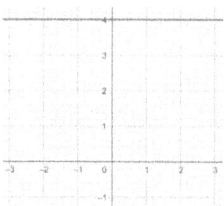

Lesson 31: Graphing Lines Using Intercepts

It is often very important to know where a line hits the x-axis or y-axis. Where a line hits the x-axis is called the x-intercept. Where a line hits the y-axis is called the y-intercept. Here is the graph of the line $y = 2x + 2$.

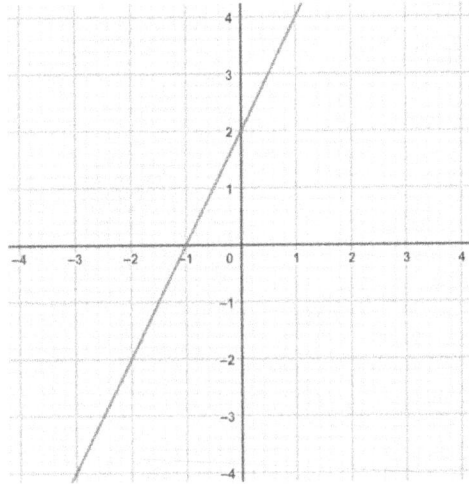

Give the value of the intercepts. (They must be given as ordered pairs, not just a single number.)

a)   x-intercept: $(\_\_, 0)$

y-intercept: $(0, \_\_)$

The x-intercept always occurs when the y-coordinate is zero. The y-intercept always occurs when the x-coordinate is zero.

Try another. Don't forget to give the intercepts as ordered pairs.

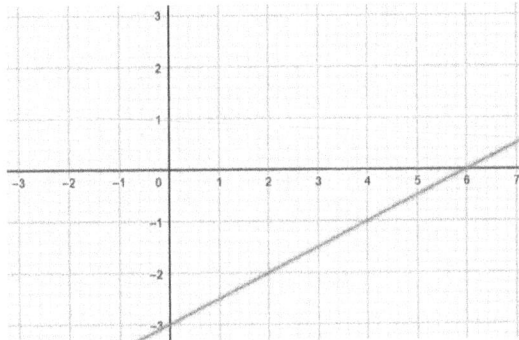

b)   x-intercept:

y-intercept:

We have been taking the intercepts off the graph. However, we can find them by entering zeros into the equation. Since the x-intercept occurs when $y = 0$, we can enter a 0 into the equation for y. Likewise, the y-intercept occurs when $x = 0$, and so we can enter a 0 in for x. (Normally, if we are graphing a line, we pick three points, but if you have both the intercepts that is sufficient.)

c) Find the intercepts for the line $2y + 3x = 6$.

	x	y
x-intercept		0
y-intercept	0	

d) Plot the intercepts and graph the line.

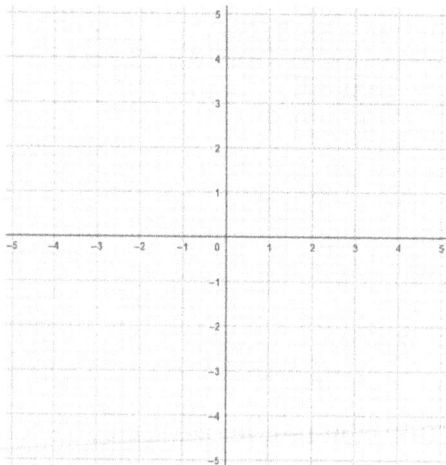

e) Graph one more line by using the intercepts.

$$5x - 2y = 10$$

	x	y
x-intercept		0
y-intercept	0	

f) Plot the intercepts and graph the line.

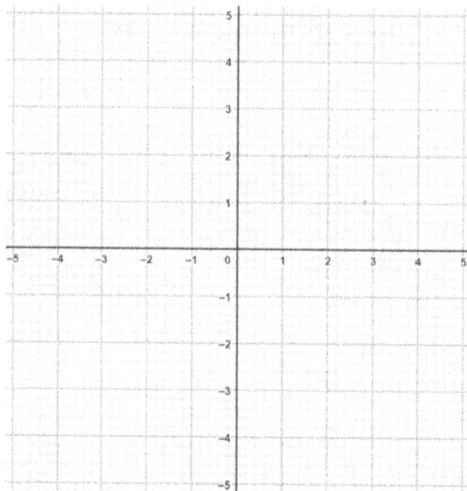

Lesson Thirty-One: Additional Exercises

Here is the graph of the relation $5x + 2y = 10$. Give the values of the x and y intercepts.

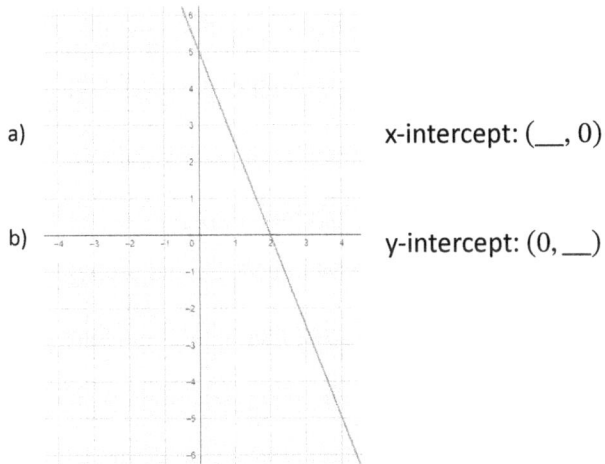

a)

b)

x-intercept: (___, 0)

y-intercept: (0, ___)

Here is the graph of the line $y = -\frac{2}{3}x - 2$. Give the values of the x and y intercepts.

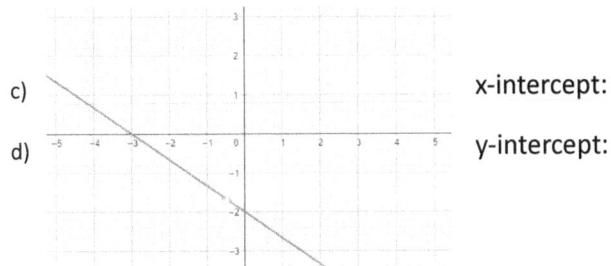

c)

d)

x-intercept:

y-intercept:

Find the intercepts for the line $4x - 2y = -12$.

	x	y
x-intercept		0
y-intercept	0	

e)

f)

Plot the intercepts and graph the line.

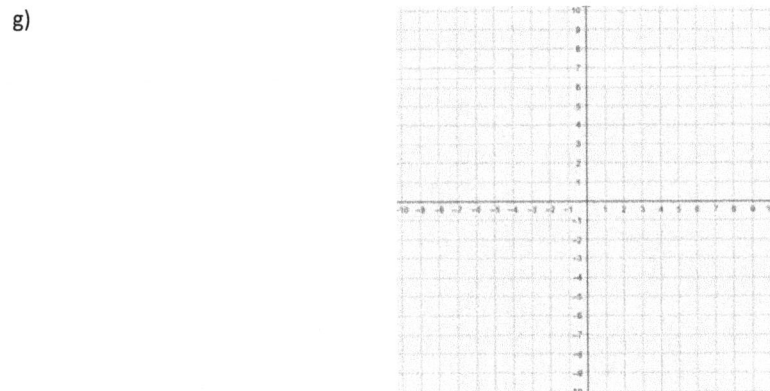

g)

Find the intercepts for the line $y = \frac{1}{3}x + 2$.

h)

i)

	x	y
x-intercept		0
y-intercept	0	

Plot the intercepts and graph the line.

j)

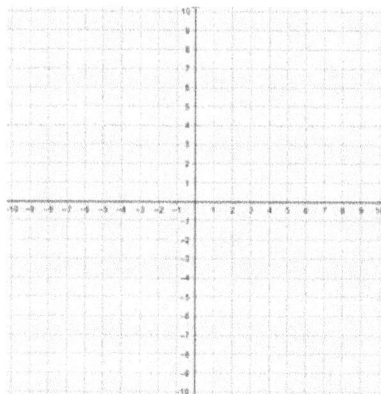

Lesson Thirty-One: Answers

a) x-intercept: (-1, 0)

y-intercept: (0, 2)

b) x-intercept: (6, 0)

y-intercept: (0, -3)

c) x-intercept: (2, 0)

y-intercept: (0, 3)

d)

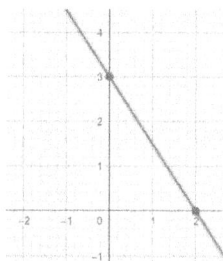

e) x-intercept: (2, 0)

y-intercept: (0, -5)

f)

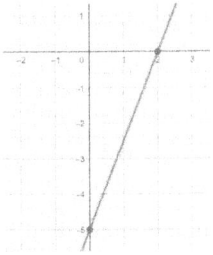

Lesson Thirty-One: Additional Exercises Answers

a) $(2,0)$

b) $(0,5)$

c) $(-3,0)$

d) $(0,-2)$

e) $(-3,0)$

f) $(0,6)$

g)

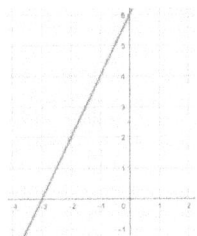

h) $(-6,0)$

i) $(0,2)$

j)

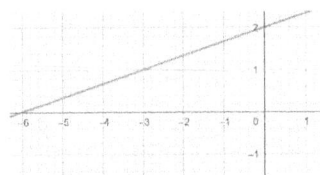

Below is the graph of the line: $y = \frac{2}{3}x - 2$.

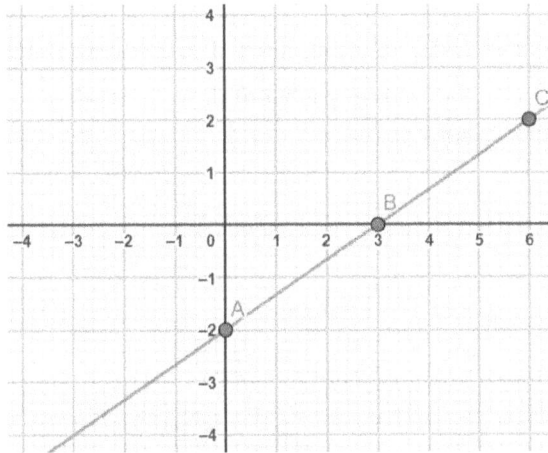

Pretend the grid lines are city block and you are standing at point C. You are going to walk up and then to the right to get to point B.

a) How many blocks up:

How many blocks to the right:

Now you are at point B. You are going to walk up and then to the right to get to point A.

b) How many blocks up:

How many blocks to the right:

c) Here is the formula again: $y = \frac{2}{3}x - 2$. Something in the formula matches with the blocks that you walked between points, what is it?

Up and down is the y direction. Up is positive. Down is negative. Right and left is the x direction. Right is positive. Left is negative.

d) Start at point A and walk down and then to the left to get to point B. (Remember both of these directions are negative.)

How many blocks down:

How many blocks left:

e) Here's the formula again: $y = \frac{2}{3}x - 2$. The number in front of the x is called the slope. It tells us how to get between any two points on the line. When you went down and left your numbers were negative. Explain why this still gets you the same slope.

Next, look at the graph of the line $y = -\frac{3}{4}x + 1$.

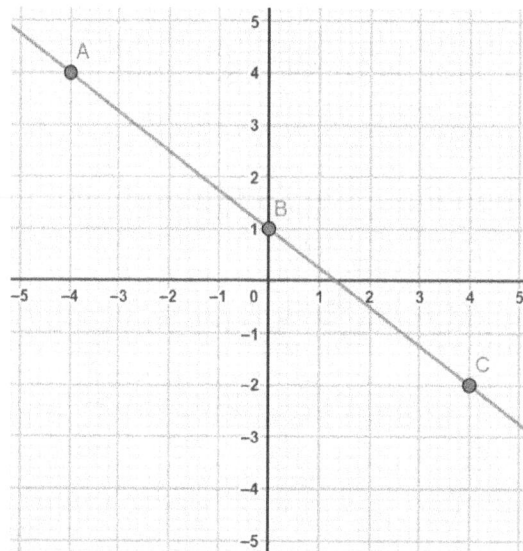

I'm choosing points that are clearly at intersections of the grid. Start at point A and go to point B.

f) How many steps down? (Remember this is negative.)

g) How many steps to the right? (Remember this is positive.)

h) Is this consistent with the slope in the equation?

This time, start at point B and go back to point A.

i)   How many steps up? (Remember this is positive.)

j)   How many steps to the left? (Remember this is negative.)

k)   Below, write both the slope when we went between point A and point B and the slope when we went between point B and point A.

l)   Are they the same mathematically?

Now, start at point A and go to point C.

m)   How many steps down:

n)   How many steps to the right:

o)   What is the slope?

p)   Show why this is the same as the slope between points A and B.

Graphing works just like reading. We read from left to right and graphs do the same. A positive slope, goes uphill from left to right. A negative slope goes downhill from left to right.

Find the slope of the following graphs. (Use two points that are clearly at the intersection of the grids.) Then, put a circle around any lines with slopes that are positive and a box around any lines with slopes that are negative.

q)

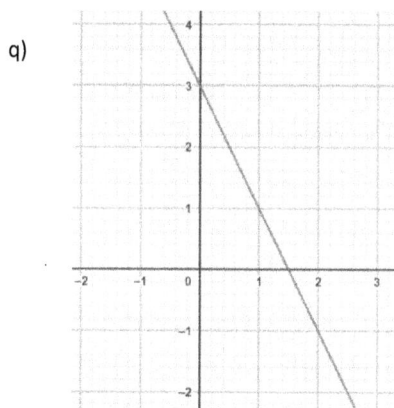

r)

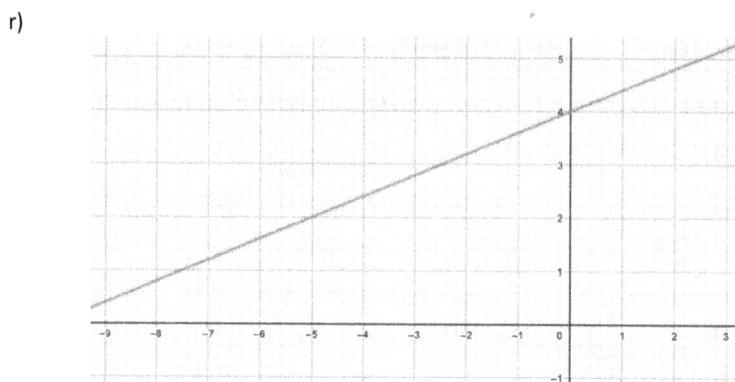

s)

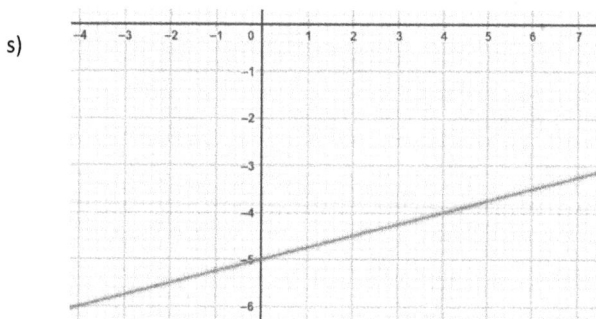

t)

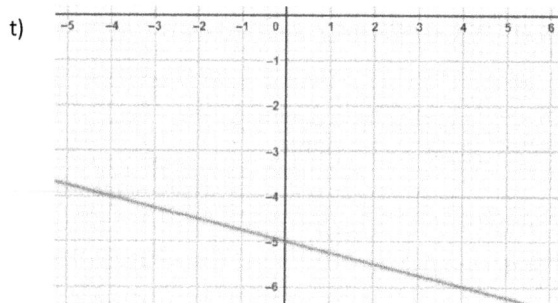

u) Here are four formulas for lines. Circle any that have positive slopes. Put boxes around any that have negative slopes.

$$y = -\frac{1}{5}x + 2$$

$$y = \frac{3}{2}x - 1$$

$$y = \frac{5}{7}x + 6$$

$$y = -2x + 3$$

Graphing this last line $y = -2x + 3$ seems like a problem. The slope is only -2. However, the solution is easy. Any number can become a fraction by putting it over the number 1. So, it becomes:

$$y = -\frac{2}{1}x + 3$$

v)   Graph it below. I've given you a starting point. Use the slope to add two points and then make your line.

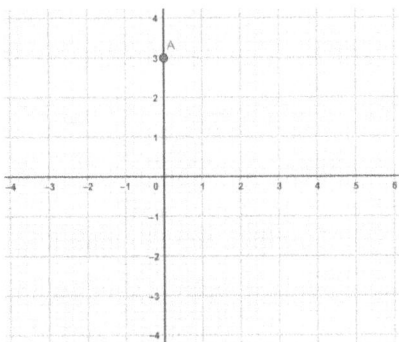

In math language, the slope of a line is called m. And it is easy to find from a graph. But if we didn't have the graph, we can find it from knowing two points on the line, using this formula:

$$m = \frac{y_2 - y_1}{x_2 - x_1}$$

This formula is nothing more than what we've been doing.

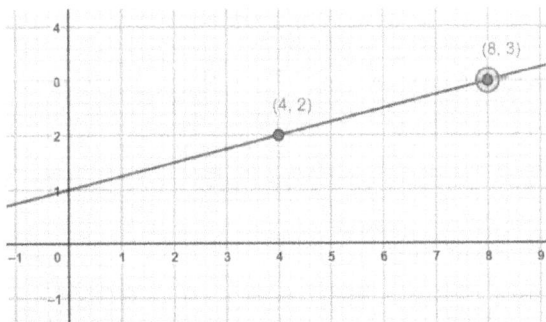

w)   From the graph, what is the slope of this line?

x)   Using the points, subtract the two y-values.

y)   Using the points, subtract the two x-values.

z)   Put your answer for the y's over the answer for the x's and you have the slope. (It doesn't matter which point you think of as the first point and which you think of as the second. Just stay consistent. Use the same point as both x_2 and y_2.

Use the formula $m = \frac{y_2 - y_1}{x_2 - x_1}$ to find the slope between the following points:

aa) $(2, 7)$ and $(4, 10)$

bb) $(5, 3)$ and $(8, 1)$

On this next one, be careful, subtracting a negative number makes two negatives in a row, which changes it to addition.

cc) $(-4, 5)$ and $(-2, 3)$

dd) Below I have graphed the horizontal line $y = 3$. Use the formula and the two points I have marked to find the slope.

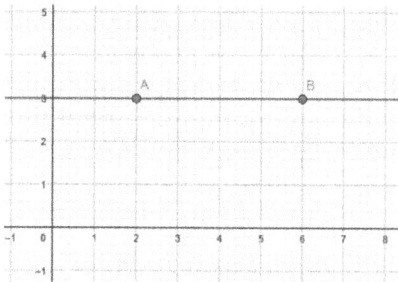

ee) Horizontal lines always have a slope of zero. Why?

ff) Here is a graph of the line $x = 2$. Use the formula and the two points I have marked to find the slope.

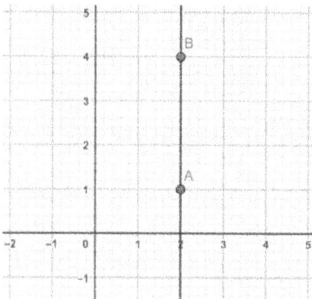

gg) Vertical lines always have an undefined slope. Why?

Lesson Thirty-Two: Additional Exercises

a) Use the graph to find the slope of the line.

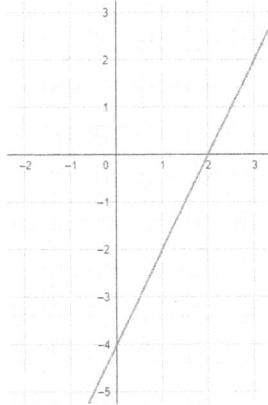

b) Use the graph to find the slope of the line.

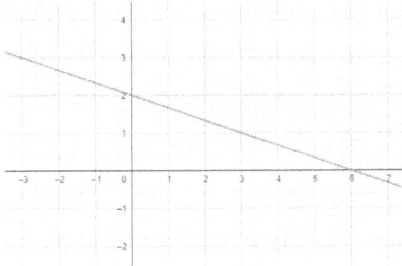

Use the formula $m = \frac{y_2 - y_1}{x_2 - x_1}$ to find the slope between the following points:

c) $(2, 4)$ and $(6, 10)$

d) $(5, 10)$ and $(3, 6)$

e) $(-3, 4)$ and $(-1, -4)$

f) $(-1, 7)$ and $(3, 7)$

g)   Give the slope of the following line.

$m =$

h)   Give the slope of the following line.

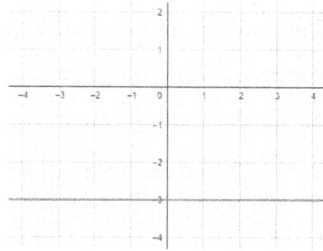

$m =$

Lesson Thirty-Two: Answers

a) 2 up; 3 right

b) 2 up; 3 right

c) the slope $\frac{2}{3}$

d) 2 down; 3 left

e) the negatives cancel

f) down 3

g) right 4

h) yes

i) up 3

j) left 4

k) $\frac{-3}{4}, \frac{3}{-4}$

l) yes

m) down 6

n) right 8

o) $-\frac{6}{8}$

p) $-\frac{6}{8} = -\frac{3}{4}$

q) $-\frac{2}{1}$

r) $\frac{2}{5}$

s) $\frac{1}{4}$

t) $-\frac{1}{4}$

u) positive: $y = \frac{3}{2}x - 1$; $y = \frac{5}{7}x + 6$

   negative: $y = -\frac{1}{5}x + 2$; $y = -2x + 3$

v)

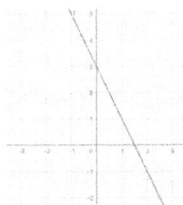

w) $\frac{1}{4}$

x) 1

y) 4

z) $\frac{1}{4}$

aa) $\frac{3}{2}$

bb) $-\frac{2}{3}$

cc) $\frac{-2}{2} = -1$

dd) 0

ee) the y values are always the same

ff) undefined

gg) because the x coordinates are in the denominator and they are always the same.

Lesson Thirty-Two: Additional Exercises Answers

a) $m = 2$

b) $m = -\frac{1}{3}$

c) $m = \frac{3}{2}$

d) $m = 2$

e) $m = -4$

f) $m = 0$

g) undefined slope

h) $m = 0$

Relations and Functions

Lesson 33: Graphing a Line Using the Slope and a Point

a) If we know the slope of a line then we only need one point on that line to graph it. I've given you a point, graph the line if $m = \frac{2}{3}$.

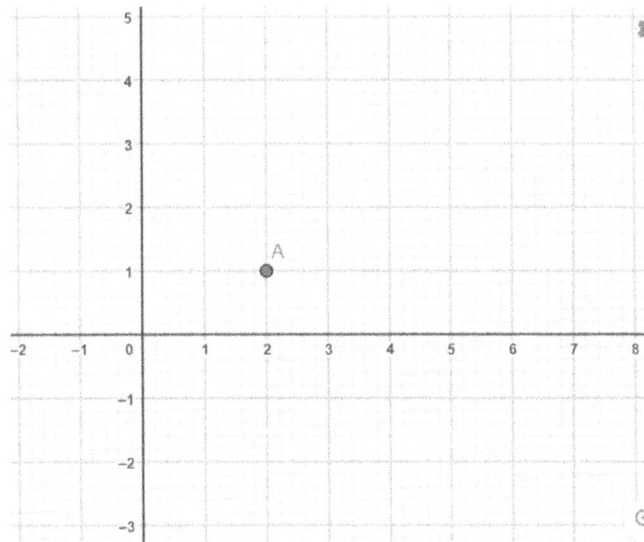

b) Graph the line given the following information:

$m = -\frac{2}{3}$ and containing the point $(4, 1)$

(To graph a negative slope, give the negative either to the top $\frac{-2}{3}$ or to the bottom $\frac{2}{-3}$. They are the same thing. Don't give the negative to both the top and the bottom. That wouldn't be the same.)

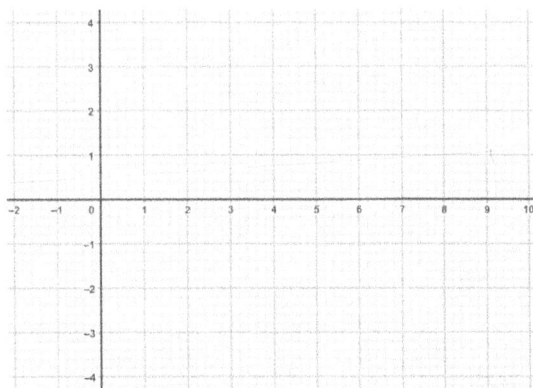

We can find the line with the slope and any point. However, as we will see, the most common point to use is the y-intercept. (Remember, the y-intercept is where the line hits the y-axis.)

c)   Graph the line with the slope $m = \dfrac{1}{2}$

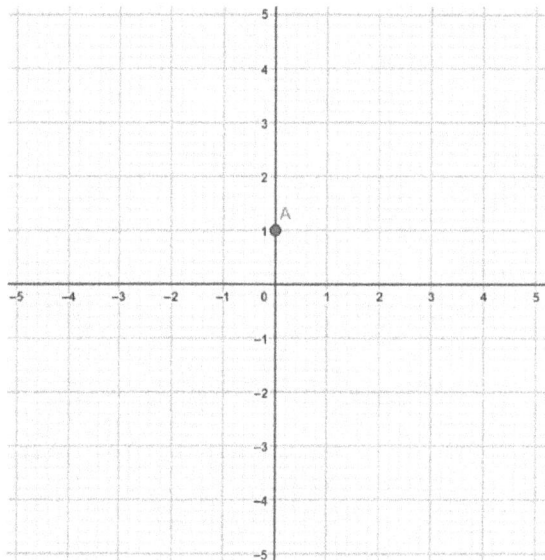

d)   Graph the line with the given information:

$m = -\dfrac{1}{3}$ and y-intercept: $(0, 4)$

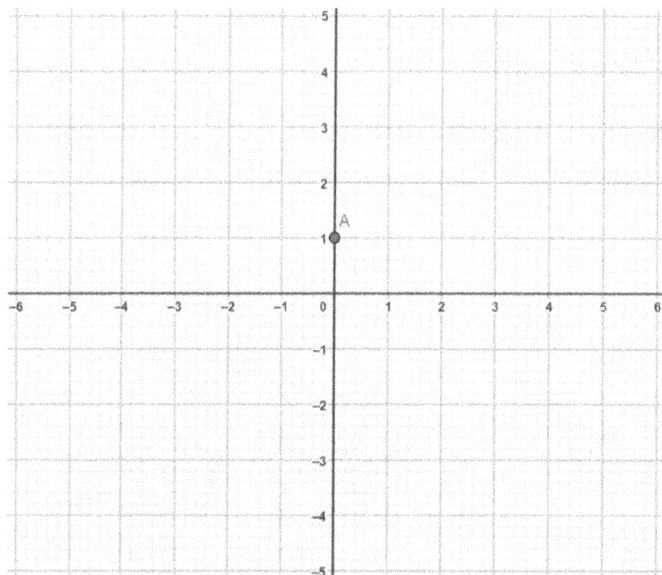

e) When a slope isn't a fraction, make it a fraction by putting it over 1.

$m = 3$ and y-intercept: $(0, -2)$ (Make this slope $m = \frac{3}{1}$)

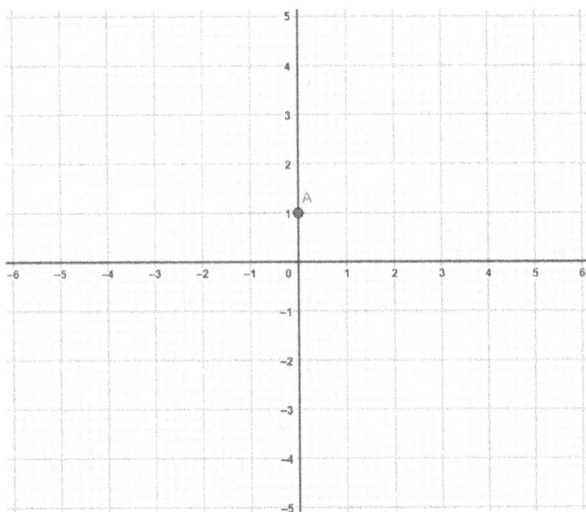

f) Try one more, but this one has a twist.

$m = 0$ and y-intercept: $(0, 1)$

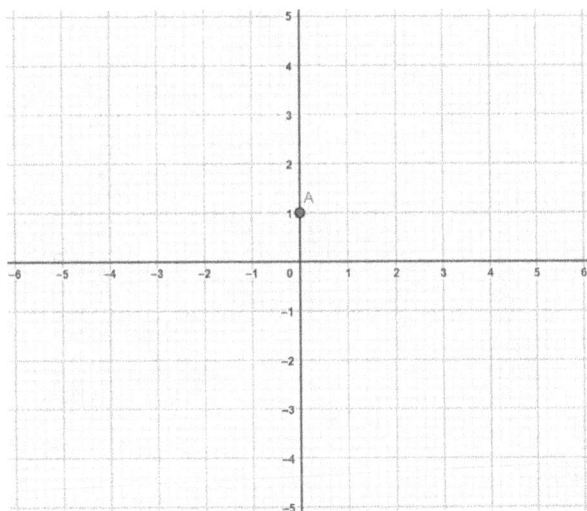

Lesson Thirty-Three: Additional Exercises

a)   Graph the line with the given information:

$m = \frac{1}{2}$ and y-intercept: $(0, 2)$

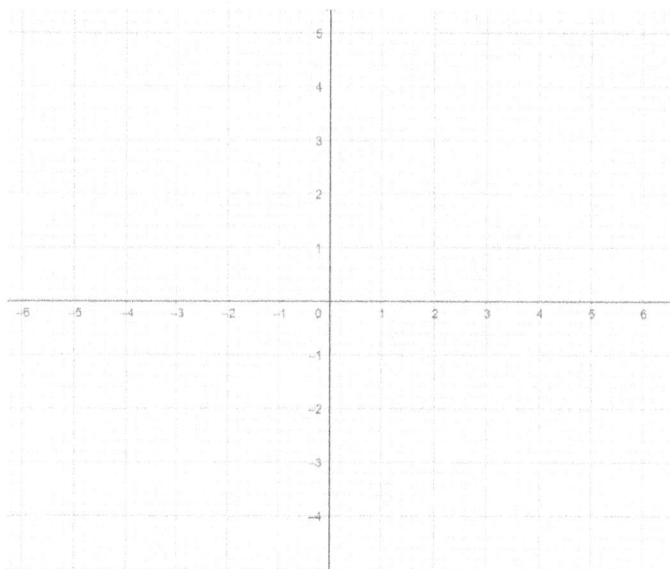

b)   Graph the line with the given information:

$m = -\frac{2}{3}$ and y-intercept: $(0, 5)$

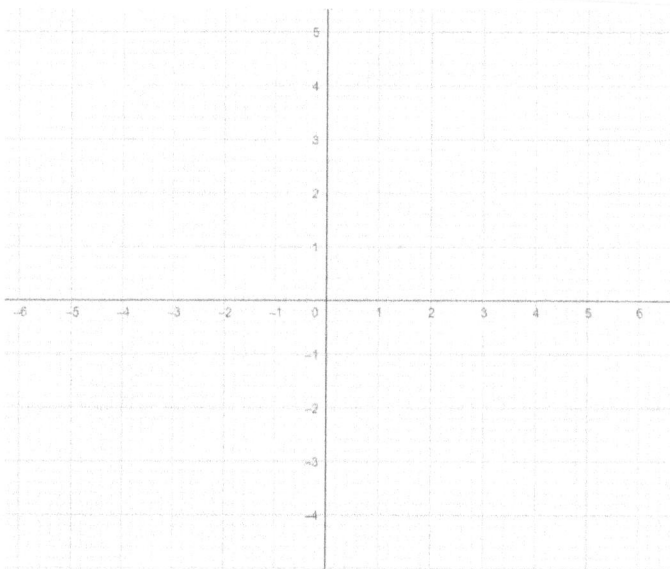

c) Graph the line with the given information:

$m = 3$ and y-intercept: $(0, -4)$

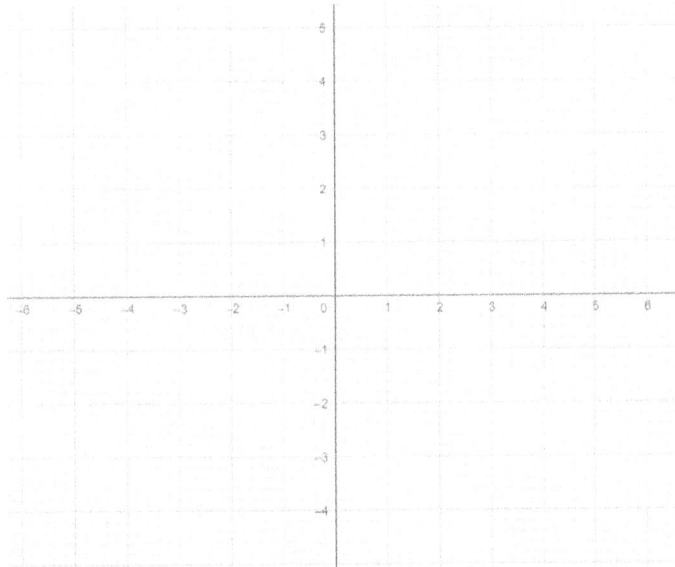

d) Graph the line with the given information:

$m = 0$ and y-intercept: $(0, 1)$

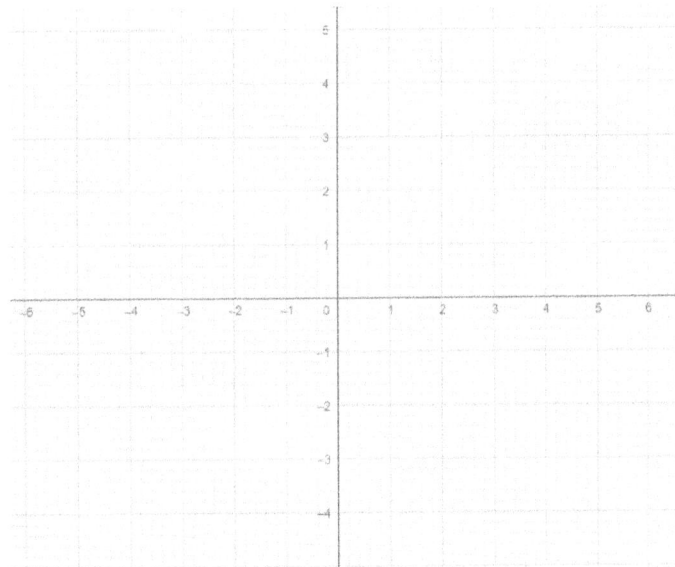

Lesson Thirty-Three: Answers

a)

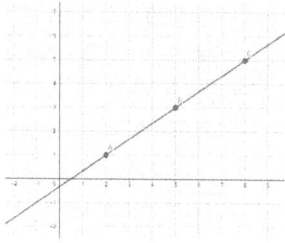

b)

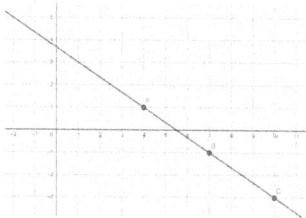

c)

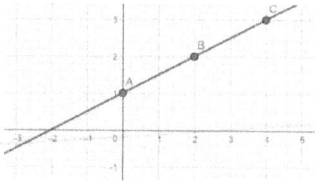

d)

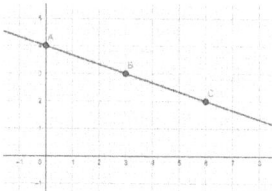

e)

f)

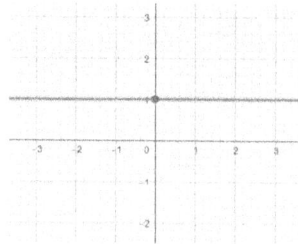

Lesson Thirty-Three: Additional Exercises Answers

a)

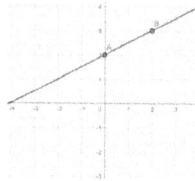

b)

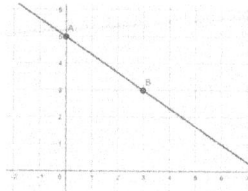

c)

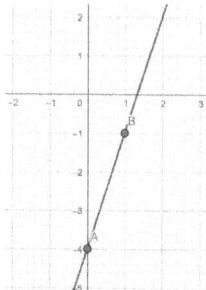

d)

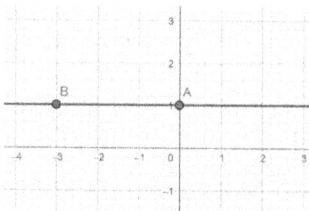

Find the y-intercept of the following lines:

(Remember, the y-intercept occurs when $x = 0$)

$$y = 3x + 5$$

a) y-intercept: $(0, \underline{\quad})$

$$y = -\frac{2}{5}x + 1$$

b) y-intercept: $(0, \underline{\quad})$

$$y = -\frac{54}{55}x - 3$$

c) y-intercept: $(0, \underline{\quad})$

d) When the equation for a line is in this form there is a connection with the y-intercept. What is the connection?

This form is called slope-intercept form $y = mx + b$. It tells you everything you need to graph the line. You have the slope (m), the value in front of the x, and the value of the y-intercept, called b.

e)   State the slope and the y-intercept for the following lines:

$$y = -\frac{3}{2}x + 5$$

Slope:

y-intercept:

Use the slope and y-intercept to graph the line.

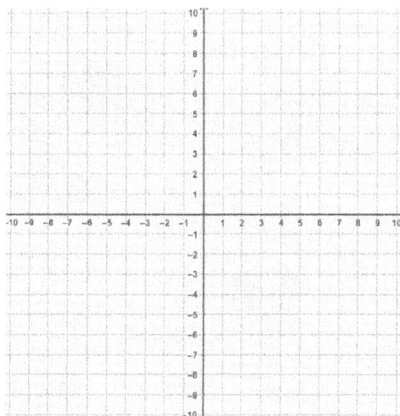

$$y = \frac{4}{5}x - 4$$

f)

Slope:

y-intercept:

Use the slope and y-intercept to graph the line.

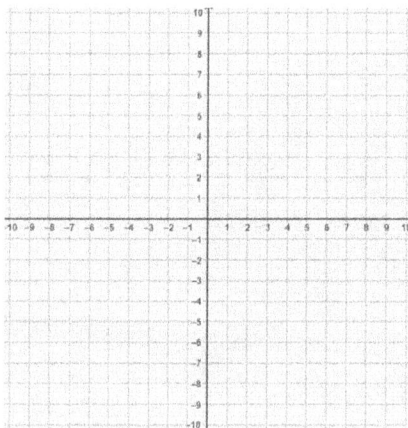

g)
$$y = -\frac{1}{3}x + 7$$

Slope:

y-intercept:

Use the slope and y-intercept to graph the line.

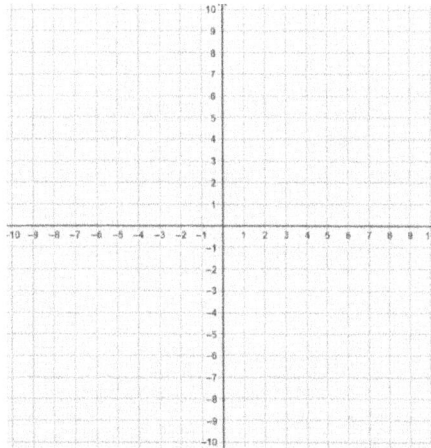

This version of the equation for a line is so helpful that we often want to rearrange an equation into this form. Use your algebra skills to put the following lines in slope-intercept form. (Get y alone on one side.)

h) $y - 2x = 3$ (Add $2x$ to both sides of the equation.)

i) $y - 5x = 8$

j) $y + 3x = 2$

When there is something in front of the y it gets a little more complicated. Let me show you.

$2y - 4x = 8$

Send the $4x$ to the other side.

$2y = 4x + 8$

Now, we divide the 2 away from both sides.

$$\frac{2y}{2} = \frac{4x + 8}{2}$$

$$y = \frac{4x + 8}{2}$$

There are two different terms on the right side. The 2 gets divided into both.

$y = 2x + 4$

k)   Try one:

$$3y + 6x = 9$$

Sometimes you try to divide but you can't go into both terms.

$$y = \frac{5x + 10}{2}$$

That's okay. It makes the slope.

$$y = \frac{5}{2}x + 5$$

l)   Try one:

$3y + 2x = 12$

248

m) Try another. It is in a bit of a different order, but the ideas are the same. When you have converted it into slope intercept form, graph the line.

$$4x + 3y = 6$$

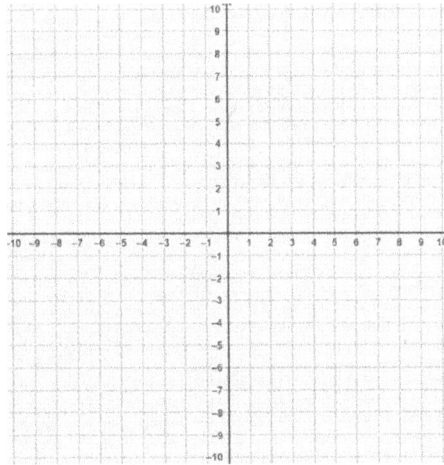

n) One this final problem, you will need to divide both sides by a -2. But everything works the same.

$$3x - 2y = 4$$

Graph the line:

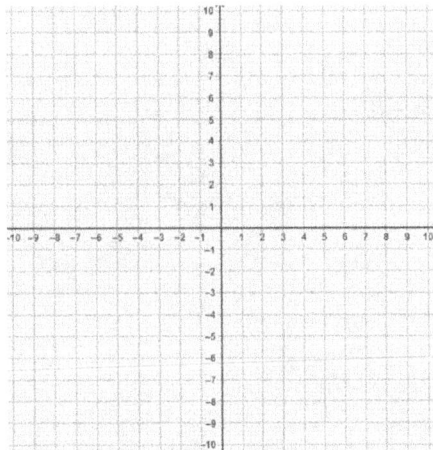

Lesson Thirty-Four: Additional Exercises

a) Graph the following line: $y = 3x + 1$.

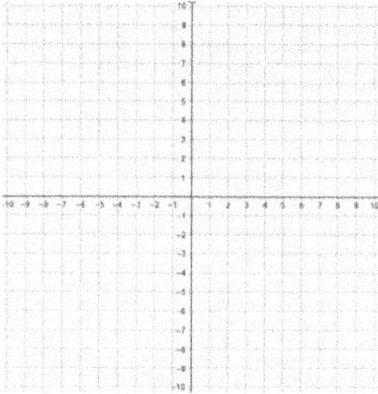

b) Graph the following line: $y = -\frac{1}{2}x + 4$

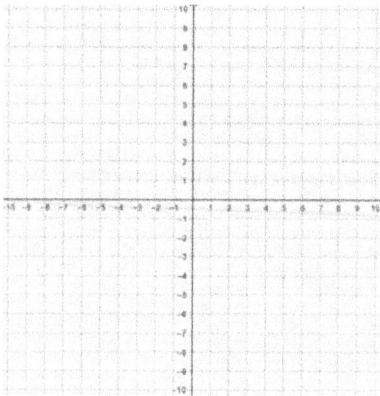

c) Graph the following line: $5x + 2y = -8$

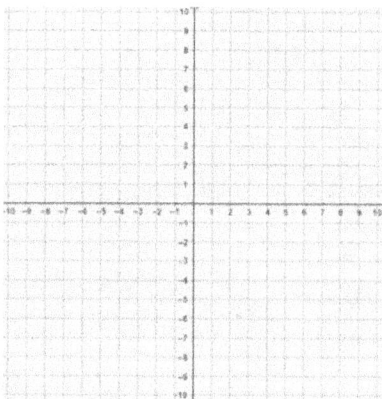

a) (0, 5)

b) (0, 1)

c) (0. -3)

d) The number alone is the intercept.

e) $-\frac{3}{2}$; (0, 5)

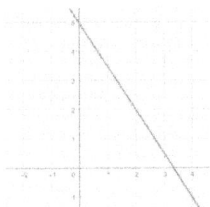

f) $\frac{4}{5}$; (0, -4)

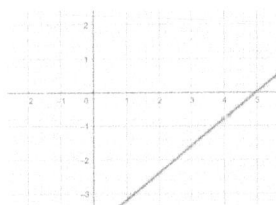

g) $-\frac{1}{3}$; (0, 7)

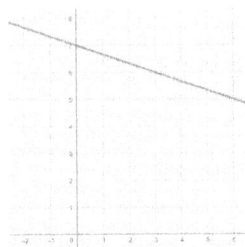

h) $y = 2x + 3$

i) $y = 5x + 8$

j) $y = -3x + 2$

k) $y = -2x + 3$

l) $y = -\frac{2}{3}x + 4$

m) $y = -\dfrac{4}{3}x + 2$

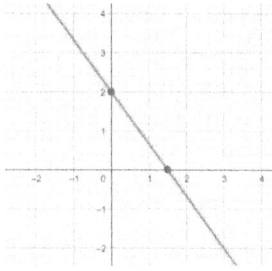

n) $y = \dfrac{3}{2}x - 2$

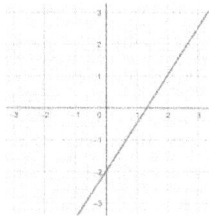

Lesson Thirty-Four: Additional Exercises Answers

a)

b)

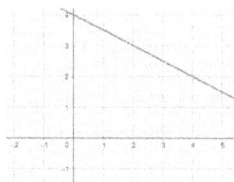

c)

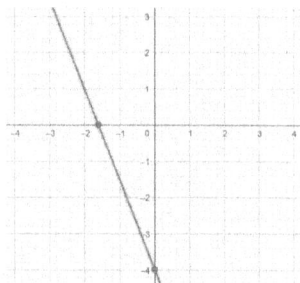

Relations and Functions

Lesson 35: Choosing the Best Method to Graph a Line

As we've looked at lines, we've seen two forms of the equation. The most important form was the slope-intercept, which looks like this:

$$y = -\frac{2}{3}x + 3$$

This form is important because it easily tells us the slope and y-intercept. Give the slope and y-intercept for this line.

a) Slope: b) y-intercept:

c) Use the slope and y-intercept to graph the equation.

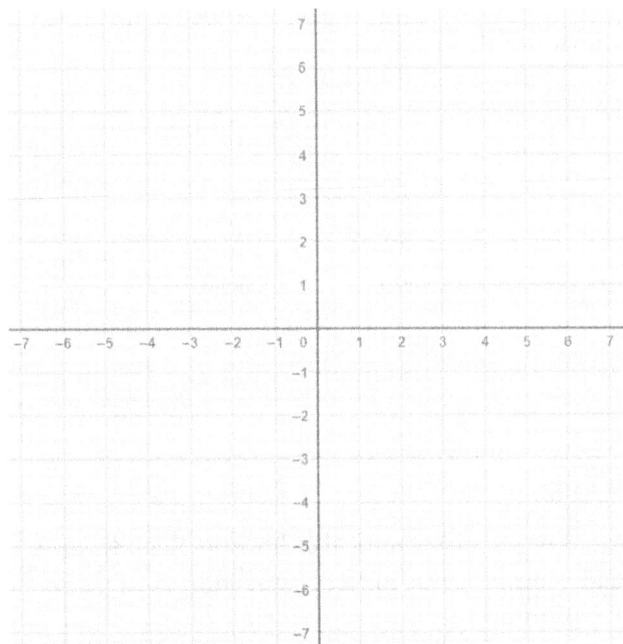

The second form of the equation for a line was the standard form. It looks like this:

$$2x + 4y = 12$$

253

The standard form gives us an easier way to compute the x and y intercepts. Find the intercepts for this line.

d) x-intercept:

e) y-intercept:

We usually use three points to graph a line, but if we have the intercepts, two points are sufficient.

f) Graph the line below, using the intercepts.

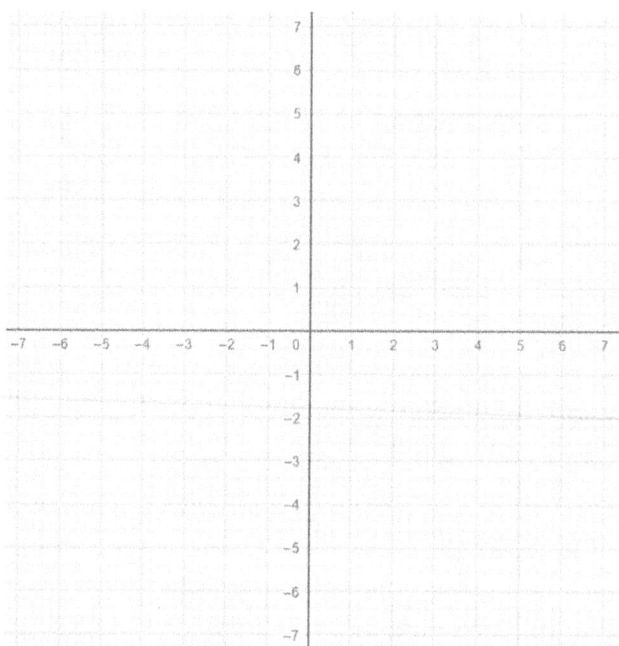

We also encountered equations which made vertical or horizontal lines. Look at the following and indicate if the equation would make a vertical or horizontal line.

g) $x = -2$ vertical or horizontal

h) $y = 5$ vertical or horizontal

Graph each of the lines below.

i) $x = -2$

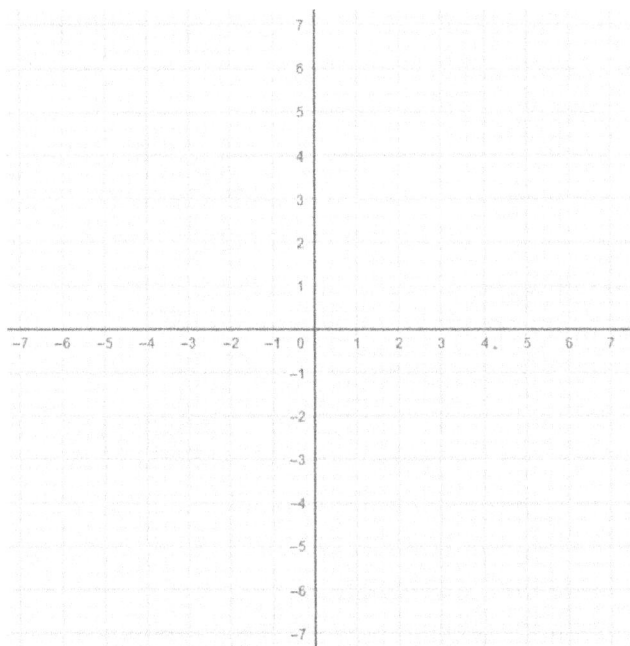

j) $y = 5$

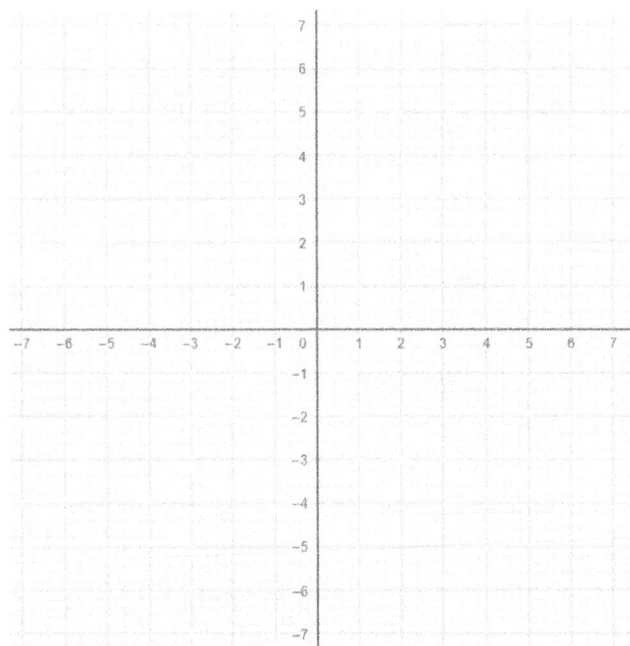

Graph the following lines:

a) $y = -1$

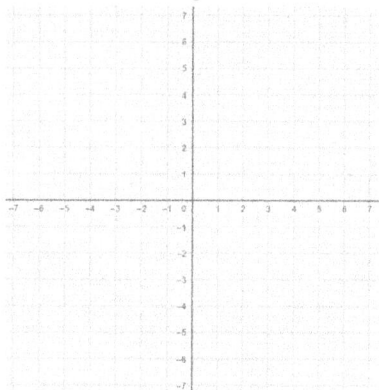

b) $x + 2y = 4$

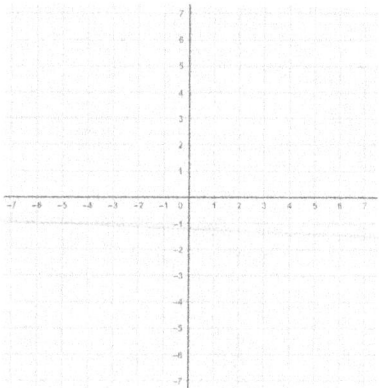

c) $y = \frac{2}{3}x - 2$

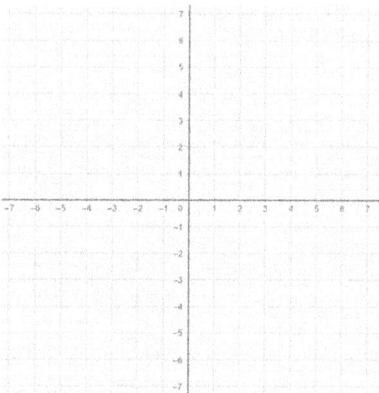

a) $-\frac{2}{3}$

b) (0, 3)

c)

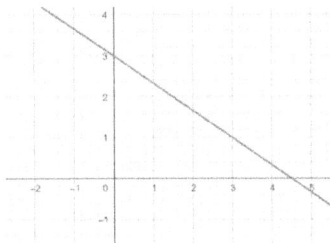

d) (6, 0)

e) (0, 3)

f)

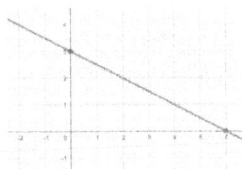

g) vertical

h) horizontal

i)

j)

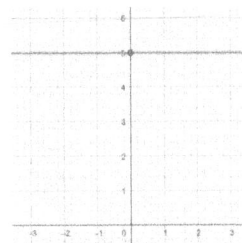

a)

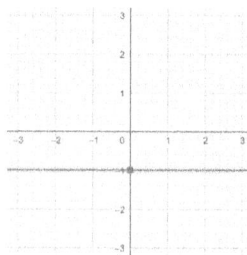

b)

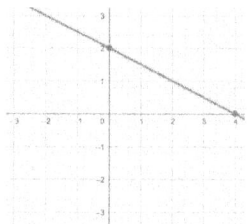

c)

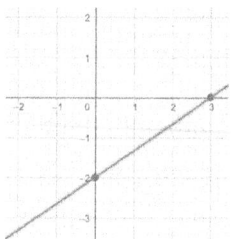

We want to look at pairs of lines which have special relationships: parallel lines and perpendicular lines. Parallel lines are like railroad tracks. They are always an equal distance apart and they will never cross. Perpendicular lines make a perfect x, creating $90°$ angles between the lines.

a) Using slope-intercept form, graph both of these lines on the same graph:

$$y = \frac{3}{2}x + 5$$

$$y = \frac{3}{2}x + 1$$

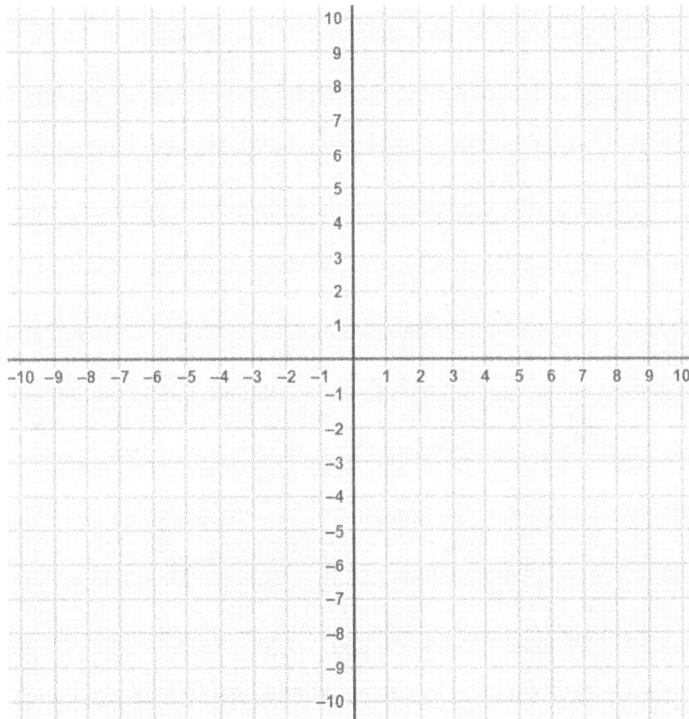

b) Are these parallel lines or perpendicular lines?

c) What in the formula creates this kind of pair?

d) Graph the following lines:

$$y = -\frac{2}{3}x - 1$$

$$y = \frac{3}{2}x + 2$$

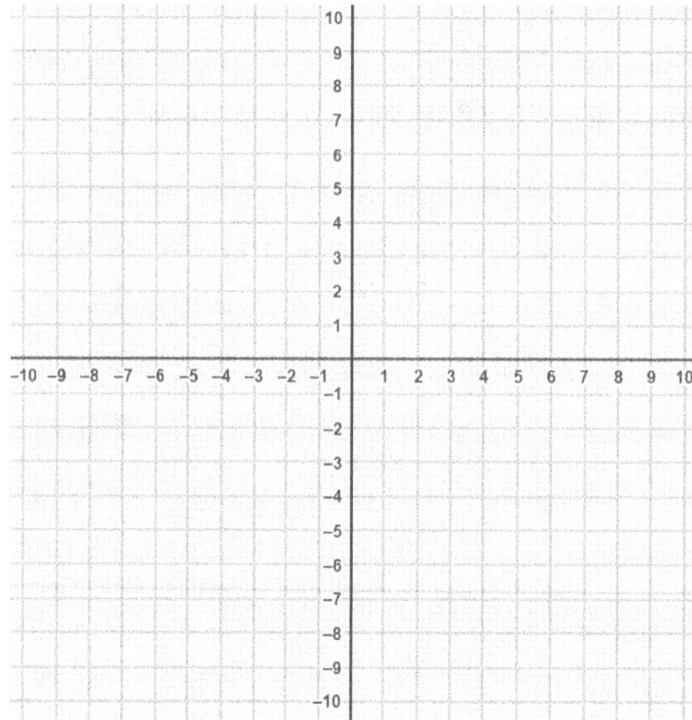

e) Are these lines parallel or perpendicular?

f) Two things about the slope are causing this. What are those two things?

Put the following pairs of equations into slope-intercept form, and indicate if they are parallel, perpendicular, or neither. (Remember, the key is the slope. Keep in mind what you learned above.)

$$7x + 2y = 3$$

$$2x + 7y = 5$$

g)   Parallel, Perpendicular, or Neither.

$$3x - 2y = 6$$

$$y = \frac{3}{2}x + 1$$

h)   Parallel, Perpendicular, or Neither.

$$4y + 3x = 12$$

$$3y - 4x = 15$$

i)   Parallel, Perpendicular, or Neither.

Lesson Thirty-Six: Additional Exercises

Indicate if the following pairs of lines are parallel, perpendicular or neither:

a) $y = \frac{4}{5}x - 2; y = -\frac{5}{4}x + 3$

b) $y = -\frac{1}{3}x + 12; y = \frac{1}{3}x - 7$

c) $y = \frac{2}{7}x - 1; -2x + 7y = 14$

d) $3x + 2y = 10; y = \frac{2}{3}x + 5$

e) $4x + 5y = 5; y = -\frac{5}{4}x + 2$

Lesson Thirty-Six: Answers

a)

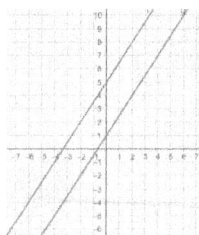

b) parallel

c) the same slope

d)

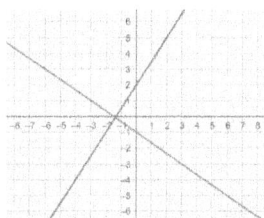

e) perpendicular

f) inverse slopes (upside down) with the opposite sign

g) neither

h) parallel

i) perpendicular

Lesson Thirty-Six: Additional Exercises Answers

a) perpendicular

b) neither

c) parallel

d) perpendicular

e) neither

Find the slope and y-intercept of this line.

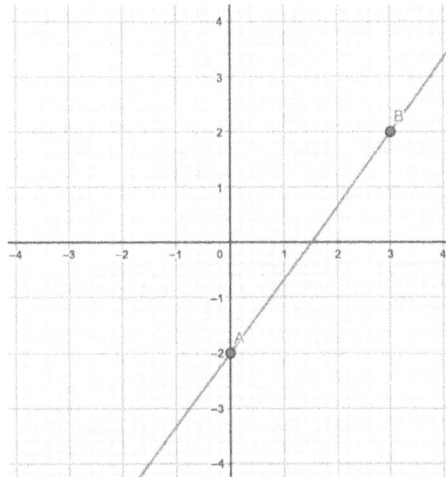

a) Slope: $m =$

b) Y-intercept: $(0, \underline{})$

c) If we have the slope of the line and the y-intercept, we can simply put those facts together to make the slope-intercept equation for the line. What is the equation of the line above?

d) We don't need to have the y-intercept to find the equation of the line. If we have the slope and any point, we can still find it. Here's how. What is the name of the formula I've written below? (It has been written backward, but it is still something you know.)

$$\frac{y - y_1}{x - x_1} = m$$

I've now rearranged the formula by multiplying the denominator up to the other side.

$$y - y_1 = m(x - x_1)$$

This is called the point-slope form. I know the slope and one point (x_1, y_1). I don't know a second point, so I'm just going to leave them as the variables (x, y). Here is the graph of the same line we already found, but I've marked a new point.

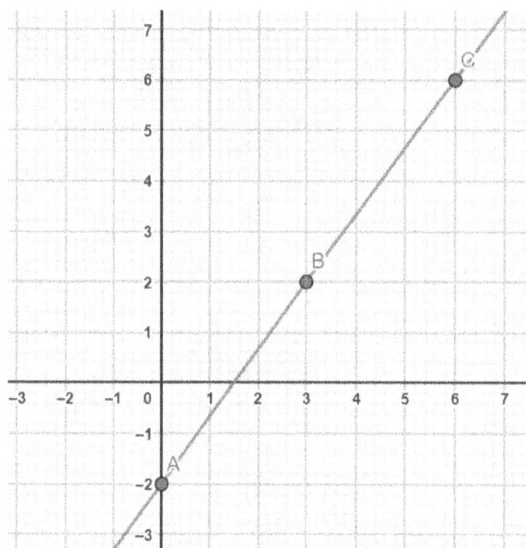

e)   What are the coordinates of the new point C?

f)   What was the slope of this line?

g)   Put the slope and point C into the point-slope equation. Point C is (x_1, y_1).

$$y - y_1 = m(x - x_1)$$

h)   Finally, put the equation into slope-intercept form. (You must first distribute the m and then send y_1 to the other side. Combine any like terms.)

i)   Is your equation the same as the one you originally found above?

j) Find an equation of the line using the point-slope formula. (You must then put it into slope-intercept form.)

$$m = \frac{2}{3}$$

$(9, 2)$

k) Try another.

$$m = -\frac{1}{3}$$

$(6, -4)$

You could take the idea one step further and find the equation for the line if you only had two points. First, you would need to find the slope. Then, use the slope and either of your points to do a problem like we did above. Once again, let's use the same graph we started with.

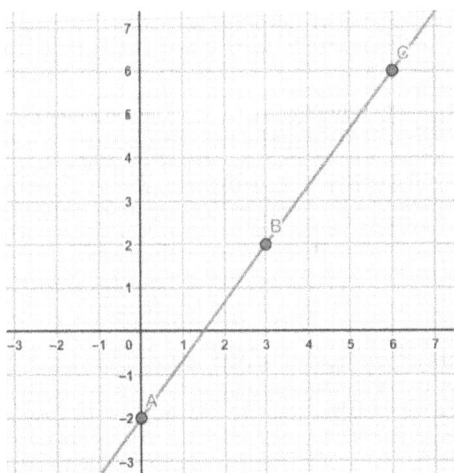

l) What are the coordinates of point C?

m) What are the coordinates of point B?

n) Find the slope between the two points by using the slope formula?

$$m = \frac{y_2 - y_1}{x_2 - x_1}$$

o)   Now, put the slope and point C into the point-slope formula. Then, convert the equation of the line into slope-intercept form.

p)   Next, put the slope and point B into the point-slope formula. Then, convert the equation of the line into slope-intercept form.

If you've done it correctly, you should get the same formula (the one you already knew) using either point B or point C.

q)   Try one more. Find the equation of the line given these two points.

$(3, 2)$

$(5, 6)$

r)   Let's end with a couple tricks. Find the equation of the line given these two points.

$(5, 1)$

$(5, 7)$

(Hint: you could find the slope, but notice that the two x-coordinates are the same. What kind of line would that be?)

s)   And between these two points.

$(4, -1)$

$(15, -1)$

(Hint: Notice that the two y-coordinates are the same. What kind of line would that be?)

Lesson Thirty-Seven: Additional Exercises

Find the equation for the following lines, given the slope and the point on the line:

a) $m = \frac{1}{4}$ $(8,4)$

b) $m = -\frac{2}{3}$ $(9,-10)$

c) $m = -\frac{5}{6}$ $(-18,14)$

Find the equation for the following lines, given two points.

d) $(-3,-1),(3,1)$

e) $(-2,3),(1,0)$

f) $(-7,-4),(-2,-1)$

Lesson Thirty-Seven: Answers

a) $\frac{4}{3}$

b) $(0, -2)$

c) $y = \frac{4}{3}x - 2$

d) slope formula

e) $(6, 6)$

f) $\frac{4}{3}$

g) $y - 6 = \frac{4}{3}(x - 8)$

h) $y = \frac{4}{3}x - 2$

i) yes

j) $y = \frac{2}{3}x - 4$

k) $y = -\frac{1}{3}x - 2$

l) $(6, 6)$

m) $(3, 2)$

n) $\frac{4}{3}$

o) $y = \frac{4}{3}x - 2$

p) $y = \frac{4}{3}x - 2$

q) $y = 2x - 4$

r) $x = 5$

s) $y = -1$

Lesson Thirty-Seven: Additional Exercises
Answers

a) $y = \frac{1}{4}x + 2$

b) $y = -\frac{2}{3}x - 4$

c) $y = -\frac{5}{6}x - 1$

d) $y = \frac{1}{3}x$

e) $y = -x + 1$

f) $y = \frac{3}{5}x + \frac{1}{5}$

We want to graph a line parallel to the one graphed. Point A is the y-intercept of the line which we want to graph.

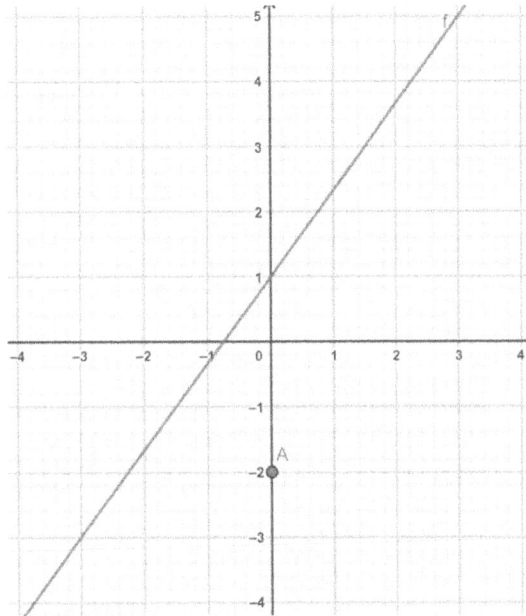

a) Find the slope of the graphed line.

b) If our line is parallel to the one already graphed, what do we know about the slope of our line?

c) What is the slope of the parallel line we want to graph?

d) Using that slope and point A, graph the parallel line on the graph above.

e) Let's do the same thing without being able to look at the slope from the graph. We want to graph a line parallel to $y = -\frac{2}{3}x + 3$ with a y-intercept of $(0, -1)$. What is the slope of the line we want to graph?

f) Graph it below.

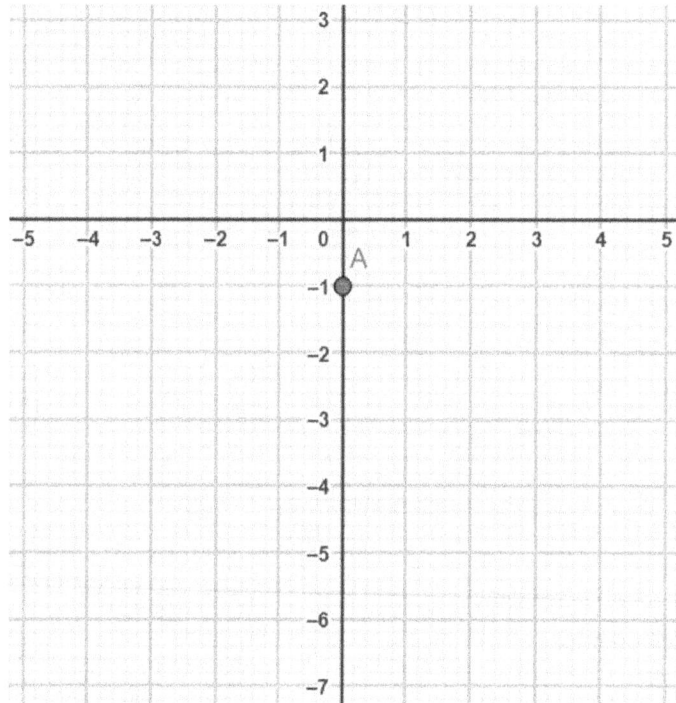

Let's do one more. This time, we aren't going to graph it; we are just going to find the equation for our new line using the point-slope formula.

Find the equation for a line parallel to the line $y = \frac{1}{2}x + 4$ and going through the point $(2, 1)$.

g) What is the slope of our line?

h) Use the point-slope formula with our slope and the point $(2, 1)$. (Rearrange it to put it into slope-intercept form.)

Next, we want to graph a line perpendicular to the one graphed. Point A is the y-intercept of the line we want to graph.

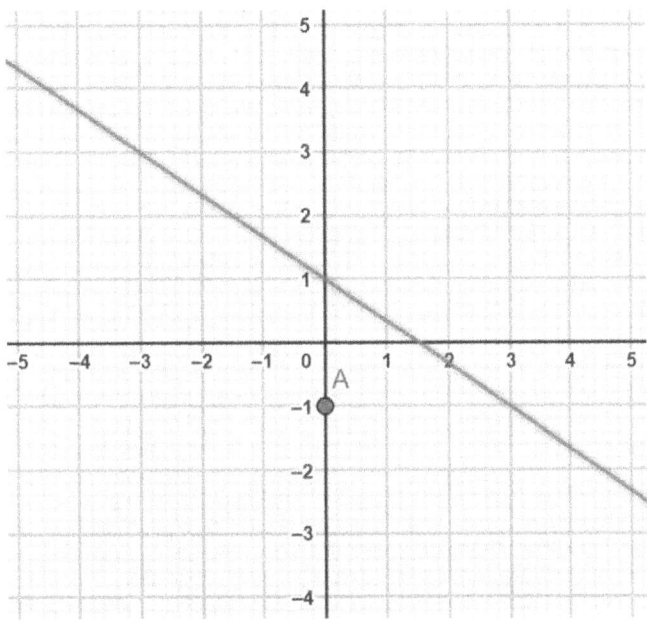

i)   Find the slope of the graphed line.

j)   If our line is perpendicular to the one already graphed, what do we know about the slope of our line?

k)   What is the slope of the perpendicular line we want to graph?

l)   Using that slope and point A, graph the perpendicular line on the graph above.

Once again, let's do the same thing without being able to take the slope from the graph.

m) We want to graph a line perpendicular to $y = -3x + 3$ with a y-intercept of $(0, -1)$. What is the slope of the line we want to graph? (Remember -3 is the same as $-\frac{3}{1}$.)

n) Graph it below.

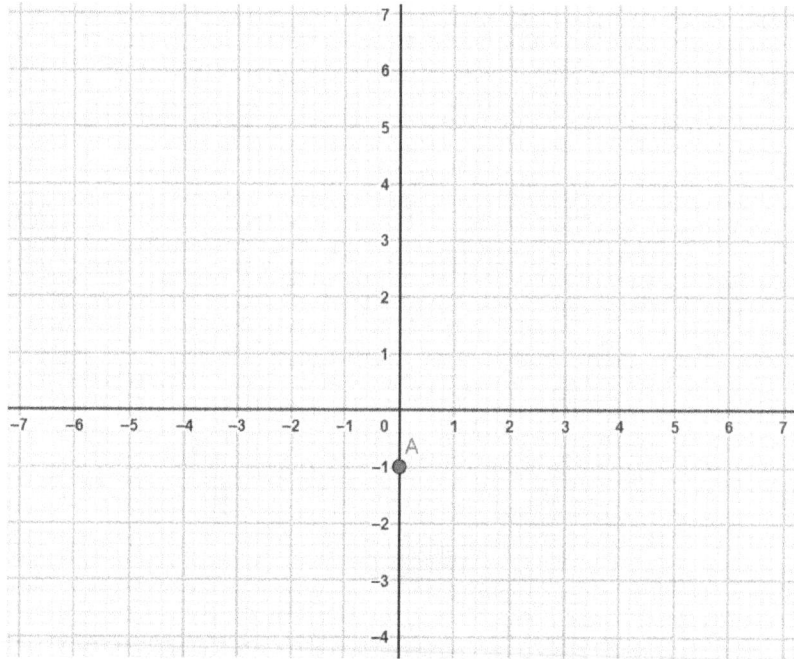

As before, let's find a perpendicular line without graphing. Find the equation for a line perpendicular to the line $y = \frac{1}{2}x + 4$ and going through point $(2, 1)$.

o) What is the slope of our line?

p) Use the point-slope formula with our slope and the point $(2, 1)$. (Rearrange it to put it into slope-intercept form.)

Lesson Thirty-Eight: Additional Exercises

a) Find the equation of the line parallel to the line $y = \frac{3}{2}x - 2$ and passing through the point $(2, 4)$.

b) Find the equation of the line perpendicular to the line $y = -\frac{1}{5}x + 3$ and passing through the point $(1, 4)$.

c) Find the equation of the line parallel to the line $4x + 5y = 10$ and passing through the point $(5, -1)$.

d) Find the equation of the line perpendicular to the line $3x - 2y = 8$ and passing through the point $(-3, -3)$.

Lesson Thirty-Eight: Answers

a) $\frac{4}{3}$

b) It is the same.

c) $\frac{4}{3}$

d)

e) $-\dfrac{2}{3}$

f)

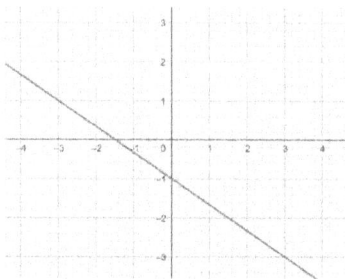

g) $\dfrac{1}{2}$

h) $y = \dfrac{1}{2}x$

i) $-\dfrac{2}{3}$

j) it is the inverse with the opposite sign

k) $\dfrac{3}{2}$

l)

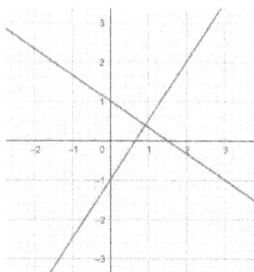

m) $\dfrac{1}{3}$

n)

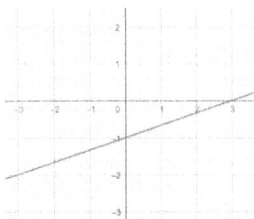

o) -2

p) $y = -2x + 5$

Lesson Thirty-Eight: Additional Exercises Answers

a) $y = \frac{3}{2}x + 1$

b) $y = 5x - 1$

c) $y = -\frac{4}{5}x + 3$

d) $y = -\frac{2}{3}x - 5$

Solve each of the following equations for the variable y.

a) $4x + 2y = 8$

b) $x^2 + y^2 = 16$

c) For the following questions, look at the graph and fill in the chart below.

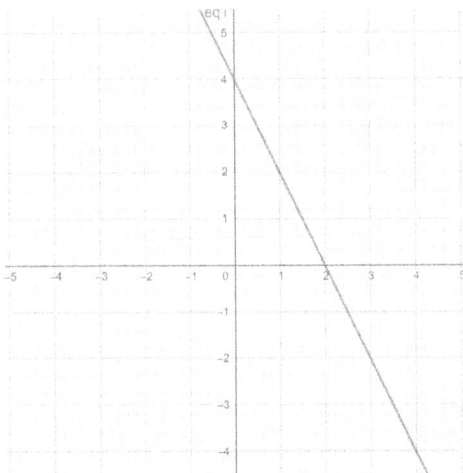

x	y
-1	
0	
1	
2	
3	

d) Fill in the circles below to show the relationship between the x's and the y's. (I've done the x circle for you. At the end of the arrow, write down the y which goes with that x.)

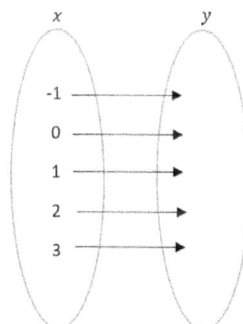

e) Fill in the chart below. (The y values won't always be whole number. That's okay. Just estimate the best you can. If an x has two partner values, put them both in the box.)

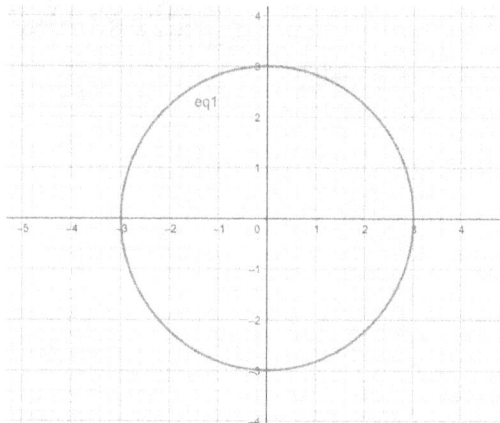

x	y
-2	
-1	
0	
1	
2	

f) Now show the relationship between the x's and y's for the circle.

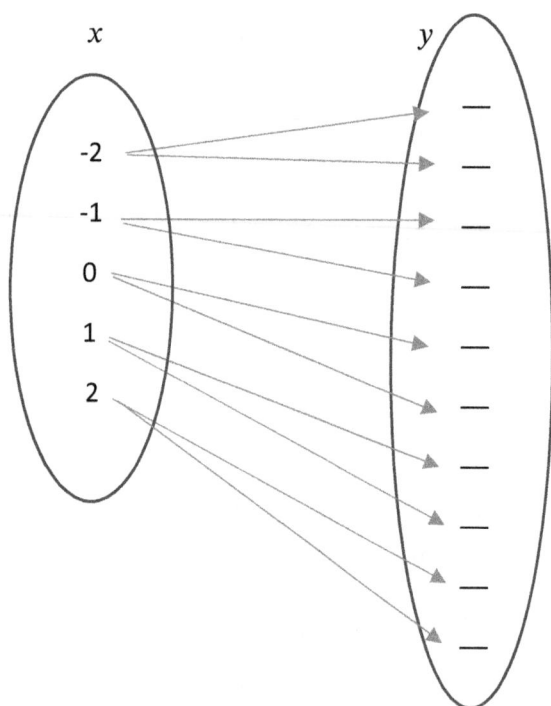

g) What is different about the relationship between the x's and y's in the first graph (the line) and in the second graph (the circle)?

h) A line is called a function (y is a function of x). A circle is not a function (y is not a function of x). The difference is in the relationship between the x's and y's. What does it take to have a function?

A function is a special kind of relationship. (In math, we just call it a relation.) First, the relation is set up like a machine, with the equation solved for y:

$$y = 3x + 2$$

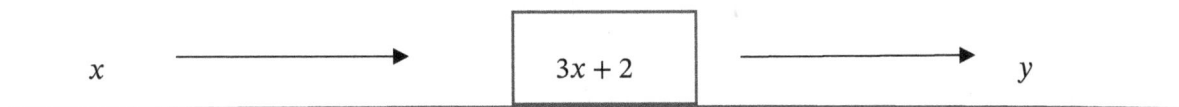

Next, x goes into the machine. Then, it gets calculated. Here, it is multiplied by 3 and then has 2 added. Finally, the transformed number comes out the other side of the machine as the value of y.

You have a function if only one value always comes out the other side of the machine. It wouldn't be a "functional" machine if you put something in one side and miraculously two things came out the other side.

A line is a function. For any value of x which you put in the machine, only one value of y will come out. However, a circle is not a function. When it is set up like a machine, you get:

$$y = \pm\sqrt{16 - x^2}$$

If you put an x into the machine, you get two values out for y. That is not a "functional" machine, and therefore, not a function.

If you have a function then we say that the "output is a function of the input." So, in the case of a line, y is a function of x.

The idea can be extended to any context. This is the price of used cars at a dealer:

Car	Price
2011 Lexus	11,999
2018 Chevy	14,999
2015 Toyota	13,999
2009 Ford	7,999

i) Is price a function of car? (In other words, if you start with the car, do you get out more than one price. If so, it is not a function. If you only get one price, it is a function. For instance, does any car on the list ever have two prices at the same time?)

Here is a menu at a local food stand:

Item	Cost
Hot Dog	2.99
Hamburger	3.99
Grilled Cheese	2.99
French Fries	1.99

j) Is cost a function of item? (In other words, if you start with the item, do you get out more than one cost. If so, it is not a function. If you only get one cost, it is a function. For instance, does any item on the list ever have two costs at the same time?)

k) Is item a function of cost? (In other words, if you start with the cost, do you get out more than one Item. If so, it is not a function. If you only get one item, it is a function. For instance, does any cost ever have two items at the same time?)

When something is a function, mathematicians wanted it to be clear and so they gave functions a special notation. This is the equation for a line:

$$y = 3x + 2$$

Lines are functions and so it gets the function notation:

$$f(x) = 3x + 2$$

Students can have a lot of trouble with this idea. But, $f(x)$ is still y. So, both of these indicate an ordered pair from the line:

$$(x, y)$$

$$(x, f(x))$$

Here is the reason that mathematicians thought this confusing notation was worth it:

$$f(2)$$

This tells us that we are set up like a machine. And that the number 2 is going in for x.

$$f(2) = 3(2) + 2$$

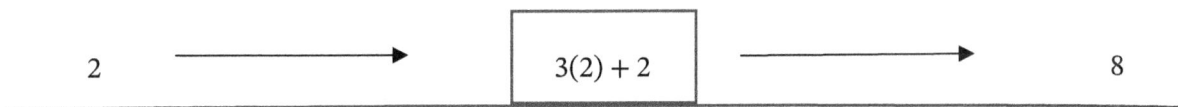

And here, we see that 2 went in for x and 8 came out the other side.

$$f(2) = 8$$

But 8 is still y, just like it always was. So, the ordered pair for this line is:

$$(2, 8)$$

Answer the following for this function:

$$f(x) = 5x - 2$$

l) $f(3) =$

m) $f(5) =$

n) $f(-2) =$

All of your answers were still values of y.

Anytime an x goes into the machine and only one value of y comes out, it is a function. This can happen for more complicated machines. Answer the following for this function:

$$f(x) = x^2 - 2x + 3$$

o) $f(1) =$

p) $f(5) =$

q) $f(-1) =$

A function doesn't have to be a story about x and y. It can be a story about anything. So, the notation can change:

$$h(p)$$

This would mean that h is a function of p. So, we could have:

$$h(p) = 3p + 2$$

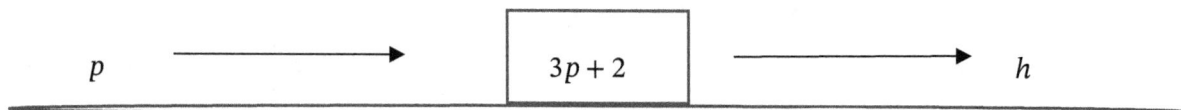

p ⟶ [$3p + 2$] ⟶ h

When we plug a value into the machine, we are evaluating the function. (Remember, to evaluate a function means to substitute that value inside of a parenthesis.) Evaluate the following function:

$$h(p) = 2p^2 - 4$$

r) $h(1) =$

s) $h(-3) =$

Finally, although it may not be obvious why this would be helpful, we can put variables inside of a function machine.

$$f(x) = x^2 - 5x$$

$$f(a) = (a)^2 - 5(a) = a^2 - 5a$$

Try a problem on your own.

$$f(x) = x^2 + 3x + 1$$

t) $f(a) =$

Lesson Thirty-Nine: Additional Exercises

Circle any of the following tables which represent a function:

a) $(2,3), (4,7), (8,11), (11,15)$

b) $(-2,-6), (-1,-4), (0,-2), (1,0), (2,2)$

c) $(1,3), (1,4), (1,5), (1,6)$

Look at the menu for the following diner.

Item	Price
Hot dog	2.99
Hamburger	3.99
Fries	1.99
Soda	1.25
Cookie	1.25

d) Is Item a function of price? Why or why not?

e) Is price a function of item? Why or why not?

Evaluate the function at the following values:

$$f(x) = 3x + 5$$

f) $f(3) =$

g) $f(-2) =$

h) $f(7) =$

Evaluate the function at the following values:

$$g(x) = x^2 + 2x - 7$$

i) $f(1) =$

j) $f(4) =$

k) $f(-3) =$

Evaluate the function at the following values:

$$h(p) = -12p + 6$$

l) $h(5) =$

m) $h(-5) =$

n) $h(-12) =$

Evaluate the function at the following values:

$$f(x) = 2x^2 - x$$

o) $f(2) =$

p) $f(-2) =$

q) $f(a) =$

Lesson Thirty-Nine: Answers

a) $y = -2x + 4$

b) $y = \pm\sqrt{16 - x^2}$

c)

x	y
-1	6
0	4
1	2
2	0
3	-2

d) At the end of each arrow is the y value from the table.

e)

x	y
-2	±2.2
-1	±2.8
0	±3
1	±2.8
2	±2.2

f) At the end of each arrow are the two y values from the table, one positive and one negative.

g) The circle has two y's for each value of x.

h) A function has one value of y for each value of x.

i) Yes, price is a function of car.

j) Yes, cost is a function of item.

k) No, item is not a function of cost.

l) 13

m) 23

n) -12

o) 2

p) 18

q) 6

r) -2

s) 14

t) $a^2 + 3a + 1$

Lesson Thirty-Nine: Additional Exercises Answers

a) function

b) function

c) not a function

d) No. The price of $1.25 has two items.

e) Yes. Each item has only one price.

f) 14

g) -1

h) 26

i) -4

j) 17

k) -4

l) -54

m) 66

n) 150

o) 6

p) 10

q) $2a^2 - a$

In our last activity, we learned that a function is a special type of relation(ship). Two variables are set-up like a machine, with an x variable being what goes into the machine, and the y variable being what comes out of the machine. If an x goes into the machine and only one value comes out the other side for y then we have a "functional" machine and therefore a function.

It turns out that having the graph of a relation makes it very easy to determine whether or not we have a function. Here is the graph of a circle. I've added a vertical line.

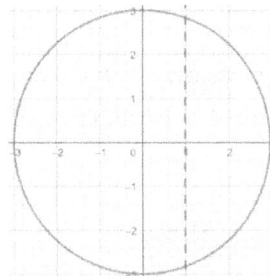

We saw previously that circles are not functions. The vertical line that I've added proves that. Vertical lines are x-lines. (The equation of the line I've added is $x = 1$.) When x is equal to 1, we hit the circle twice. That means when x is equal to 1 there are two values of y. Functions don't have two values of y for one value of x, so it isn't a function.

Look at the following graphs. Imagine drawing a vertical line. If at any point, a vertical line would hit the graph in two spots it isn't a function. If a vertical line would never hit in two spots, it is a function. Below each graph, write "function" or "not a function," depending on this vertical line test.

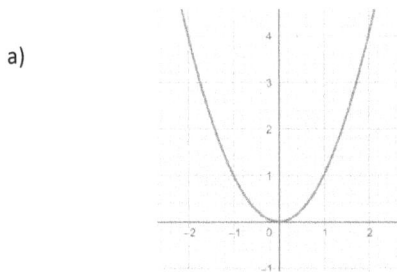

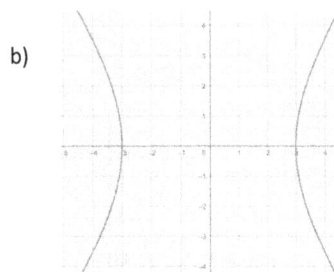

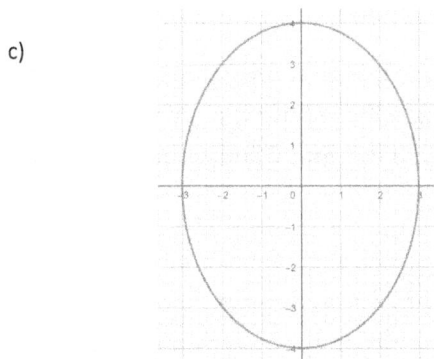

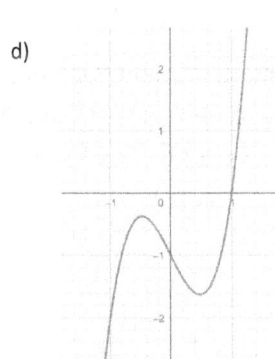

a)

b)

c)

d)

Next, we want to discuss a second kind of test called the horizontal line test. As we've explained a function, we've said when an x goes into the machine only one y comes out. Look at this function:

x	y
2	12
3	17
4	21
5	12

e) What is happening with the output when x equals 2 and when x equals 5?

The table still represents a function; each x has only one y. But mathematicians also give a name to a function where there is a unique x which leads to a unique y. They call such a function a one-to-one function.

x	y
2	12
3	17
4	21
5	12

This table is a function, but it is **not** a one-to-one function.

Look at this table:

x	y
-5	8
-2	12
0	7
3	11

This table is a function. No x leads to more than one y. But it is also a one-to-one function. No two x's ever give the same value of y. It is easy to find a one-to-one function if you have the graph. Just as a vertical line would tell us if any value of x ever had two values of y, a horizontal line would tell us if any value of y ever has two values of x.

A one-to-one function will also pass a horizontal line test. The function below is one-to-one.

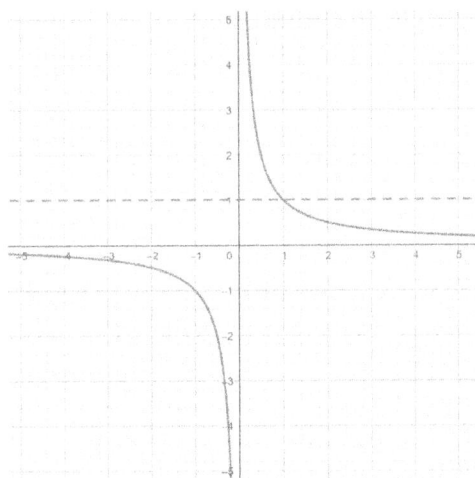

No matter where we drew a horizontal line, we will never touch the graph in two places at once. All of the following graphs are functions. Circle them if they would also pass the horizontal line test, making them one-to-one functions:

f)

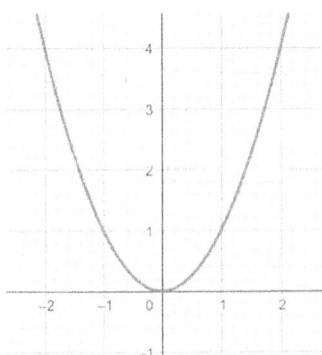

g)

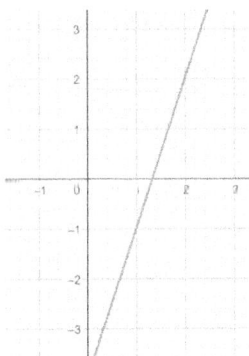

h)

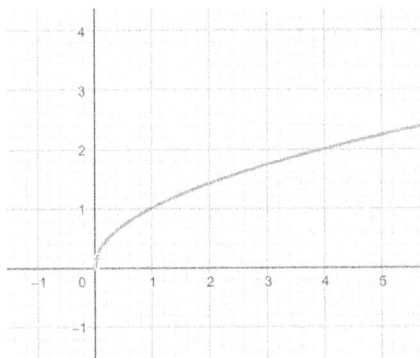

i)

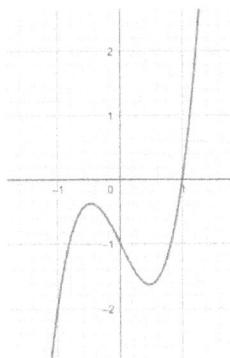

Finally, we want to read function values from a graph. It is pretty straightforward, provided you remember the notation. Let's look at the graph of a square root function. The notation $f(4)$ means that 4 was put into the machine as x. Then, we read the graph like normal.

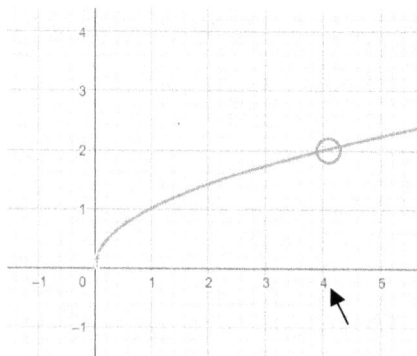

When we put in 4, we get out a y of 2. So, we have:

$$f(4) = 2$$

Evaluate the same function at 1.

j) $f(1) =$

Now, evaluate this quadratic function as indicated:

k)

$f(1) =$

l) $f(-1) =$

m) $f(2) =$

Our last type of problem asks you to go backward. You will know the y value and you must give the x value which was put into the machine to make it.

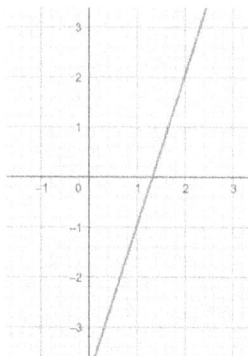

Use the graph to the left to solve the following: (I've done the first one for you.)

Solve $f(x) = -1$

The answer would be $f(1)$. 1 is the input which makes an output of -1.

n) Solve $f(x) = 2$

292

On this last problem, they have asked you to solve for the input. Be careful, there are two values of x for each problem.

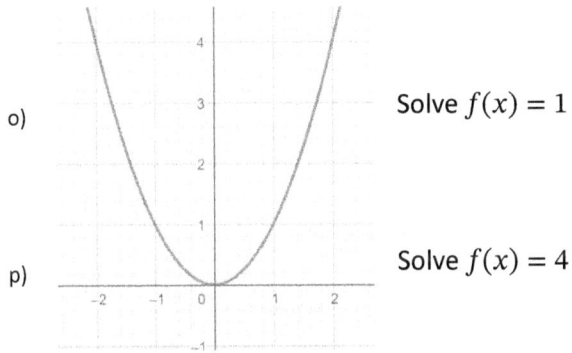

o)

Solve $f(x) = 1$

p)

Solve $f(x) = 4$

Lesson Forty: Additional Exercises

a) Use the vertical line to determine which of the following graphs are functions. Circle the functions:

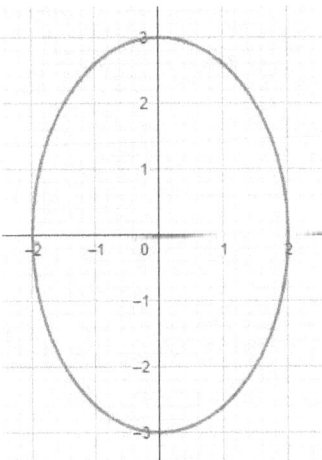

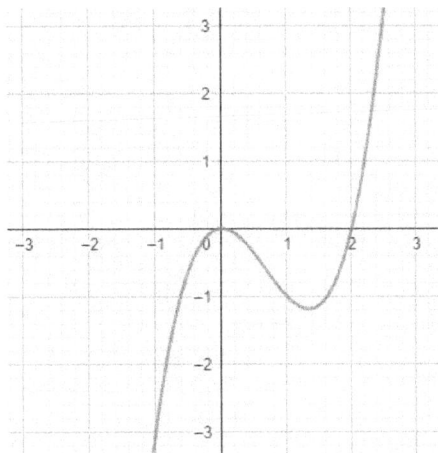

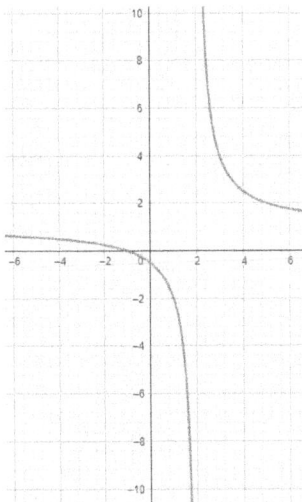

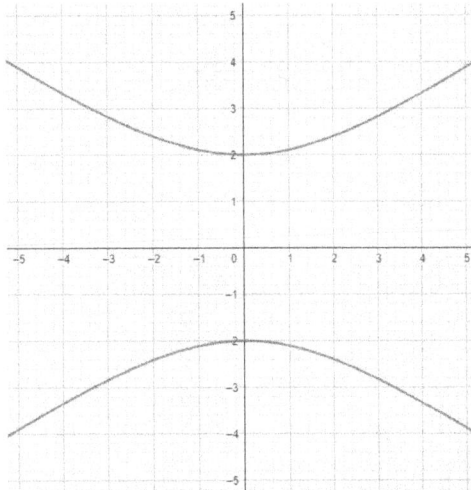

b)   One of the following functions is one-to-one, circle it:

x	y
-5	7
-2	-1
-4	10
9	3
-11	81

x	y
1	2
2	3
5	2
7	1
9	-3

x	y
-20	1
-10	0
-5	-1
0	0
5	1

c)   Use the horizontal line test to determine which of the following functions are one-to-one. Circle the one-to-one functions:

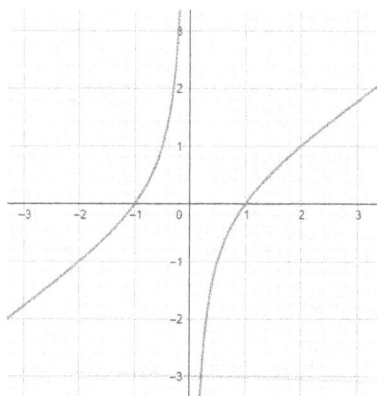

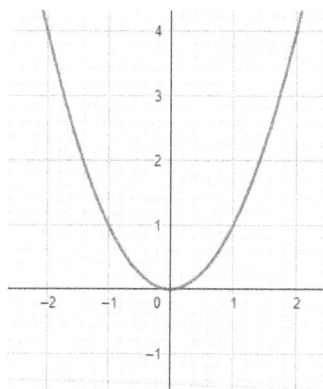

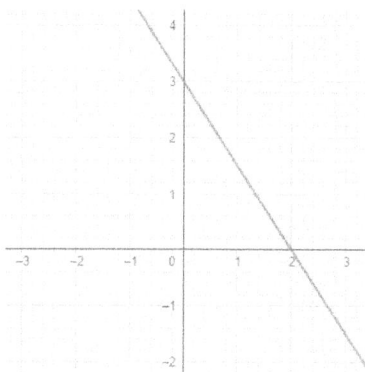

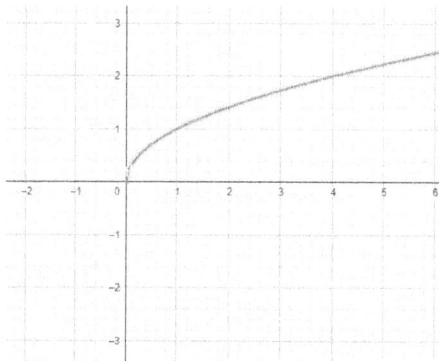

294

Find the values of the function from the graph:

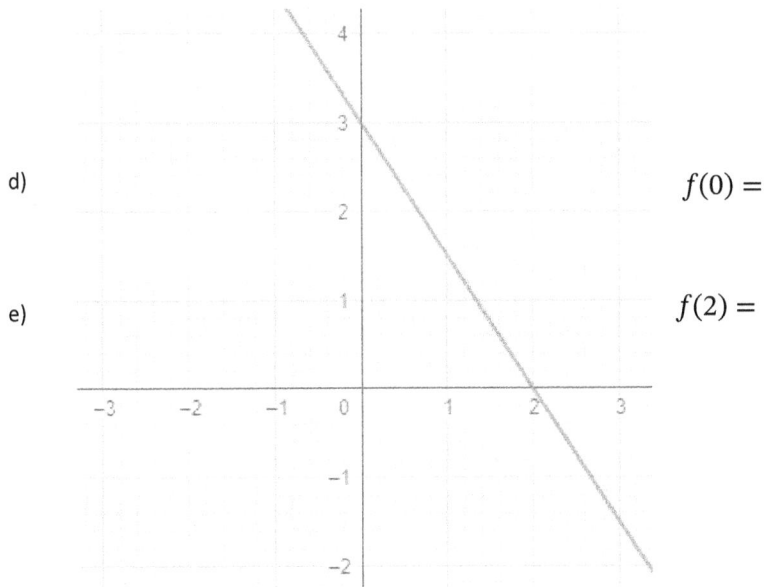

d)

e)

$f(0) =$

$f(2) =$

Find the values of the function from the graph:

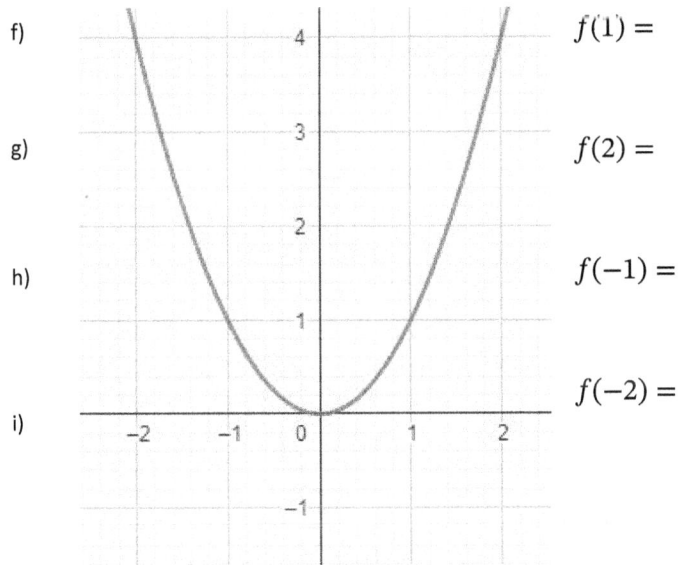

f)

g)

h)

i)

$f(1) =$

$f(2) =$

$f(-1) =$

$f(-2) =$

295

Use the graph to find the input.

j)

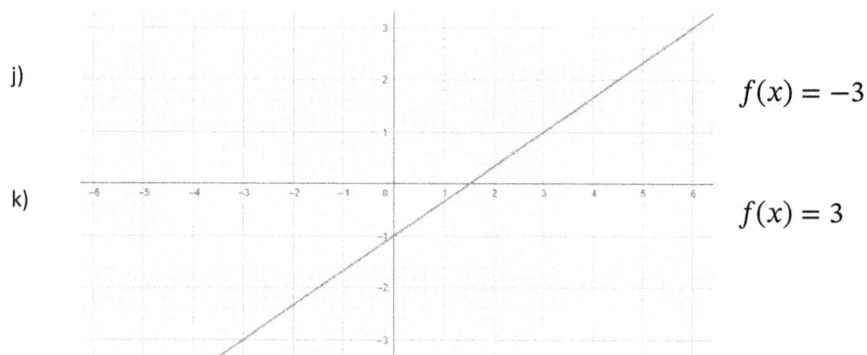

$f(x) = -3$

k)

$f(x) = 3$

Use the graph to find the input.

l)

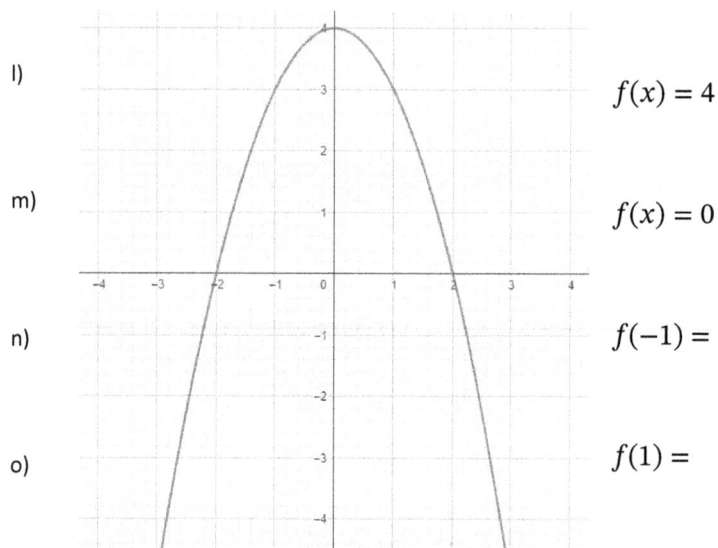

$f(x) = 4$

m)

$f(x) = 0$

n)

$f(-1) =$

o)

$f(1) =$

Lesson Forty: Answers

a) function

b) not a function

c) not a function

d) function

e) They have the same value of y.

f) Not one-to-one

g) one-to-one

h) one-to-one

i) Not one-to-one

j) $f(1) = 1$

k) $f(1) = 1$

l) $f(-1) = 1$

m) $f(2) = 4$

n) $f(2) = 2$

o) $f(1) = 1, f(-1) = 1$

p) $f(-2) = 4; f(2) = 4$

Lesson Forty: Additional Exercises Answers

a) The graphs below are functions.

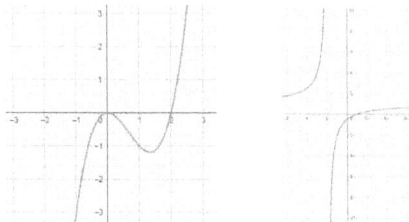

b) The following table is one-to-one.

x	y
-5	7
-2	-1
-4	10
9	3
-11	81

c) The following functions are one-to-one.

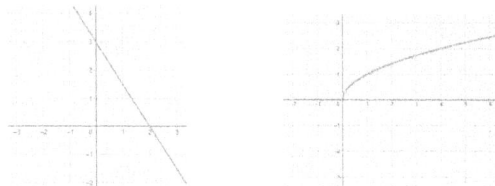

d) $f(0) = 3$

e) $f(2) = 0$

f) $f(1) = 1$

g) $f(2) = 4$

h) $f(-1) = 1$

i) $f(-2) = 4$

j) $f(-3) = -3$

k) $f(6) = 3$

l) $f(0) = 4$

m) $f(2) = 0; f(-2) = 0$

n) $f(-1) = 3$

o) $f(1) = 3$

www.ingramcontent.com/pod-product-compliance
Lightning Source LLC
Chambersburg PA
CBHW061754210326
41518CB00036B/2307